THE SILENT EPIDEMIC:
A Child Psychiatrist's Journey beyond Death Row

Understanding, Treating, and Preventing Neurodevelopmental Disorder Associated with Prenatal Alcohol Exposure

Susan D. Rich MD MPH DFAPA

Copyright © 2016 Susan D. Rich MD MPH DFAPA.

All rights reserved. No part of this book may be reproduced, stored, or transmitted by any means—whether auditory, graphic, mechanical, or electronic—without written permission of both publisher and author, except in the case of brief excerpts used in critical articles and reviews. Unauthorized reproduction of any part of this work is illegal and is punishable by law.

ISBN: 978-1-4834-4879-4 (sc)
ISBN: 978-1-4834-4881-7 (hc)
ISBN: 978-1-4834-4880-0 (e)

Library of Congress Control Number: 2016907463

Because of the dynamic nature of the Internet, any web addresses or links contained in this book may have changed since publication and may no longer be valid. The views expressed in this work are solely those of the author and do not necessarily reflect the views of the publisher, and the publisher hereby disclaims any responsibility for them.

Any people depicted in stock imagery provided by Thinkstock are models, and such images are being used for illustrative purposes only.
Certain stock imagery © Thinkstock.

Lulu Publishing Services rev. date: 6/10/2016

Table of Contents

Dedication vii
Foreword ix
Acknowledgments: xvii
Introduction xix

Part I: Illuminating the Problem 1
What Is Neurodevelopmental Disorder associated with Prenatal Alcohol Exposure? 3
   1. I Am [not] Adam Lanza's Psychiatrist 7
   2. Behind Bars 18
   3. Jacob's Ladder – Climbing the Iceberg 30
   4. The Silent Epidemic 49
   5. The American Social Caste System 70
   6. Predator…Or Prey? 85
   7. Historical Perspectives: Illuminating the Past to Enlighten the Future 103
   8. Preconception Health - A Lot Happens Before Pregnancy Recognition 122
   9. The Way Opens – Enlightenment of a Social Drinker 150

Part II: Shifting Social Paradigms 167
   10. Doctors' Responsibility in ND-PAE Prevention 169
   11. Nature versus Nurture … or Both! 180
   12. Augmenters to Foster Care 193
       ➢ A Solution or Part of the Problem? 196
       ➢ Supportive Parenting 197

    ➢ Grace Court: An Enlightened Alternative 202
13. Societal Solutions to the Silent Epidemic 206

Part III: Professional and Parent Guide to ND-PAE.....................213

Appendix A: Through the Lens of a Child Psychiatrist –
Clinical Insights into Diagnosis and Treatment 215
- Appendix A-1: Sample Developmental History 217
- Appendix A-2: Screening Checklist for
  Neurodevelopmental Disorder Associated with
  Prenatal Alcohol Exposure 220
- Appendix A-3: Explanation Checklist -
  Neurodevelopmental Disorder Associated with
  Prenatal Alcohol Exposure 221

Appendix B: Therapeutic & Learning Centers – A
Neurodevelopmental Treatment Model 223
- Appendix B-1: Emotional Regulation: Mood
  Dysregulation & Autonomic Arousal 229
- Appendix B-2: Neurocognitive: Cognitive and
  Executive Dysfunctions 238
- Appendix B-3: Social Communication: Language,
  Social Skills, & Perceptual Deficits 248
- Appendix B-4: Sensory and Motor: Sensory
  Processing, Coordination & Motor Issues 250

Appendix C: Parenting a Child with ND-PAE 254

Afterword 283

Appendix D 288

About the Author 295

Dedication

For my "little fire" Earth Keeper Aidan and my "virtuous, noble" animal lover Briana – my anchors to reality. Thank you for tolerating good-enough mothering, multitasking, and midnight writing.

Foreword

It was a true pleasure to read Dr. Rich's book – *The Silent Epidemic*. For 45 years, I have seen children and adults who have had what we used to call Minimal Brain Dysfunction in medical school in 1967. However, at the time no one knew what was causing this disorder. Despite the lack of knowing the etiology, it was clear the problem was characterized by varying levels of affect dysregulation often classed as immaturity, usually in the form of an explosive temper; problems with comprehension leading to poor frustration tolerance and impulse control. These patients also had poor memories, limited adaptation to the ordinary stresses and strains of life, and poor relationships with people. Despite seeing a preponderance of evidence of the silent epidemic existed for the last 45 years, it has only been in the last 3-4 years that I have figured out what the root cause of the problem is, and Dr. Rich illuminates this root cause in this book. Being new to the fetal alcohol dance, I am a bit shy about writing a forward to a book written by a child psychiatrist who has been on the Fetal Alcohol Spectrum Disorder (FASD) crusade since for nearly a quarter of a century. However, having been on many crusades myself (Prevention of Violence, Children Exposed to Violence, HIV Prevention, Correctional Health Care, Prevention of Head Injuries, Cultural Sensitivity, and Prevention in Psychiatry), I feel qualified to speak as a zealot for the US's Public Health.

Based on my own studies and the studies of others, I am convinced the silent epidemic in this book is America's biggest public health problem and probably dwarf's polio. Dr. Rich confirms this by suggesting that there are approximately 1,680,000 individuals with Neurodevelopmental Disorder Associated with Prenatal Alcohol Exposure (ND-PAE) born since 1973 when the problem was identified in American medical literature. However, most of the American medical community and the American public are still asleep at the wheel. To wake us up, Dr. Rich has written this book which she describes as 'not as war cry but a hearing aid to overcome deafness to public health warnings about the leading known preventable cause of birth defects and neurodevelopmental issues in the Western world: prenatal alcohol exposure.' In the Introduction, she is clear that she wrote this book to 1) highlight the magnitude of this preventable condition; 2) enlighten physicians, caregivers, child welfare advocates, and professionals that there is hope for these children; and, 3) inform childbearing age alcohol consumers to make educated choices. She accomplishes this task by being a master of metaphors, nuance, and the English language and as a result the book is easy to read and enjoyable to read.

In Part one – she defines *Neurodevelopmental Disorder associated with Prenatal Alcohol Exposure* (ND-PAE) – known by the lay public as Fetal Alcohol Spectrum Disorder (FASD). Labels and terminology is always difficult for such a ubiquitous problem as alcohol exposure in utero because it shows up in so many different places – psychiatrist's offices, correctional facilities, special education facilities, foster care facilities, family practitioner's offices, etc. As a result, everyone has a different term for the individual who is quick to become frustrated and sudden to anger, who does not seem to comprehend very well, and who has a difficult time flourishing in life. However, she makes the important point that, if a child is born with this acquired biological brain problem, and they are in an environment of support verses an environment of punishment; they do much better in life. We psychiatrists understand that behavior is multi-determined and despite the difficulty in communicating this complexity to the average person who is not familiar with psychiatric dynamics, Dr. Rich does an admirable job in making these complex issues easy to understand. She does not write in a way that I refer to as "psychobabble," which is how some psychiatrist talk to be technically correct, impress people, or because they do not know the true meaning of the concepts they are trying to explain. She writes so that seasoned psychiatrists, ordinary physicians, parents, educators, correctional officers, foster care workers, and everyone else under the sun can understand. She uses plenty of examples and historical reference to illustrate how long this problem has been right in front of us, but how we have failed to act in a responsible manner.

In chapter one (I am Not Adam Lanza's Psychiatrist), she raises the issue that I have been helped to be aware of by an educator in the crusade – Jody Allen Crowe - who wrote a book on FASD called <u>The Fatal Link</u>. I recently wrote a commentary for <u>Clinical Psychiatry News</u> on "school shooters" which highlighted the research that suggests that contagion might be a factor in 20% of the school shootings. In this commentary, I raise the question "What would be fundamentally off center with a person who got infected with the idea of going into a school killing a bunch of people and then committing suicide?" Some are answering FASD.

Having done a great deal of work in the area of violence prevention starting back in 1980, it is interesting to me how my interests have now come full circle, i.e. from violence prevention to fetal alcohol spectrum disorders as an etiologic factor in violence and back to violence prevention. Thus, Chapter two of Dr. Rich's book, on how many violent offenders often have FASD, also captivated my attention.

Behind Bars – chapter two - expounds on the prevalence of ND-PAE in corrections. As I ponder on how to address the silent epidemic, and look for points of advantage to place the fulcrum, corrections, especially juvenile corrections loom large. My understanding is that 2/3rds to 3/4ths of children in the Temporary Juvenile Detention Center in Cook County, Illinois have features of FASD. As I have tried to leverage some attention to this problem as well as put some systems in place that may ameliorate the problem both prenatally and postnatally, it was suggested to me that maybe it would be the lawyers on the public defender side that might raise this silent epidemic to a crescendo. Dr. Rich's work in forensic psychiatric cases indicates that the public defenders are quickly becoming aware of FASD as a mitigating circumstance in criminal behavior.

Chapter 3 tells the story of Jacob and in this narrative is well-crafted psychiatric journey of all the various aspects of trying to help a child with ND-PAE. There are so many interfaces patients, parents, and healers have to negotiate – hospitals, schools, the legal system, the foster care system, and the social security administration. There are also well meaning physicians who don't know what they are treating and cause iatrogenic problems with various medications that are indicated for schizophrenia, bipolar disorder, depression, anxiety, other neurodevelopmental disorders of childhood (ADHD, Autism, Intellectual Disabilities, Learning Disorders, Speech and Language Disorders, and Motor Disorders) but none of which quite provide relief for FASD. This chapter reveals Dr. Rich's devotion and dedication to her patients. As Danny Goleman says, "The act of compassion begins with full attention, just as rapport does. You have to really see the person. If you see the person, then naturally, empathy arises. If you tune into the other person, you feel with them ... You want to help them, and then that begins a compassionate act (Wisdom Quotes.com).

The Silent Epidemic does an excellent job of defining an epidemic from a seminal paper written in 1898. She grapples with the thorny problem that when FASD develops it is within the first 3 weeks of pregnancy, long before most women realize they are pregnant, thus the public admonitions of not drinking while pregnant are only helpful after the woman realizes she is pregnant – often times when it is too late. When I am diagnosing children and adults with ND-PAE, I am fond of asking mothers, who are usually still in their child's and adult child's life, when did they realized they were pregnant and were they doing any "social" drinking during that time of not knowing. More often than not, I get an affirmative answer. These women are still providing their children with the care, protection and support the child or adult child needs to do halfway OK in life, and it is always interesting to me how most mothers will tell the truth so their child can be properly diagnosed. Her analogy regarding how thalidomide with its obvious physical deformities was easier to address than alcohol with its invisible mental challenges is trenchant.

Next, in a rather chilling chapter - The American Social Caste System, Dr. Rich draws a comparison between Aldous Huxley's Brave New World, and how some communities have a greater likelihood of having alcohol flood their communities. And in next chapter, she highlights how patients with ND-PAE and often victimized by various forms of violence and they often inadvertently perpetrate violence. She beautifully illustrates these issues with case histories that add context to the clinical issues she is so well versed in describing.

Dr. Rich's exposition on the historic perspectives in Chapter seven of her wonderful book is a wealth of chronological facts that are mesmerizing. There are any number of psychiatric luminaries – e.g. Italian psychiatrist Lomboroso, American psychiatrist Healy (first Director of the Institute for Juvenile Research – Birthplace of Child Psychiatry), and others. She points out how Darwinian scientists were trying to figure out the effects of genetics and environment on trait development, and how one of the early teratologists (scientists who study birth defects) were studying the effects of prenatal alcohol on development in animals. Moreover, how one, Dr. Charles Stockard, suggested in 1924 that alcohol could be used to cleanse the "inferior" American citizens. At times, it seems to me that someone took his advice, which is why the problem of ND-PAE may be more rampant in communities where our "least valued citizens" reside. Additionally and thankfully, she puts to rest the hyperbole about cocaine babies, as the perception that somehow children born of women using cocaine would be completely out of control people was never true.

In the eighth chapter, Dr. Rich does an excellent explication of embryology and her use of language is poetic and agreeable to read. She explains it all so well! In the final chapter in Part one, she provides a careful chronology her own awareness of this issue and how it changed and shaped her life to be on the crusade she has undertaken.

In part two of this brilliant book, Dr. Rich discusses how to shift the social paradigms necessary to lessen the public health problem of FASD. She suggests that physicians have a tremendous responsibility in ND-PAE Prevention. She tells me that if just 30% of physicians "get it," the rest will soon be close behind. In this section, she gives credit where credit is due, e.g., to SAMHSA for their efforts to get the word about FASD out into the public's awareness. In chapter ten, she lays out a great road map for physicians to begin to do tangible steps to prevent and identify the problem of ND-PAE early.

She does an admirable job of explaining and giving profuse examples about: it is not either nature or nurture that shapes humans, but it is both/ and. In her chapter on Foster Care, Dr. Rich recommends a well thought out plan for vocational and housing programs for families afflicted with a child that had ND-PAE designed to break the vicious circle of poverty, additional children with ND-PAE, homelessness, and reliance on public assistance. The final chapter of part two provides the reader the benefit of her vast clinical experience in diagnosing and treating ND-PAE. Like the other chapters, Dr. Rich has included important factoids, e.g. the National Organization on Fetal Alcohol Syndrome reports that up to 70% of children in foster care have ND-PAE. She is appropriately critical of the foster care system as it strains to care for its wards with inadequate public funding.

Part 3 is the section of the book that contains the useful and indispensable appendices. In this part of this valuable book, Dr. Rich provides the reader with the tools to begin to address the problem that she so cleverly lays out in the first two parts of the book. She advocates, and I agree, that children, parents, and systems that support children who have ND-PAE should know the diagnosis. One grave observation during my short time in knowingly assessing adolescents and adults with ND-PAE is they often tell me they have been diagnosed as having bipolar disorder, schizophrenia, and depression – this is a nearly infallible sign of ND-PAE. Unfortunately, this diagnostic confusion leads to treatment confusion and many patients have spent most of their lives not understanding what their challenge is. This lack of understanding causes them and everyone around them to go down the wrong paths often leading to dead ends. Because it is steeped in developmental milestones and a pregnancy and birth history, her Screening Checklist for Neurodevelopmental Disorder Associated with Prenatal Alcohol Exposure is an extraordinarily useful tool to help parents get a perspective of their child's issues. Her explication of a Neurodevelopmental Treatment Model highlights the multifaceted approach to treating the multiplicity of challenges causes by poisoning the most complex organ of the body – the brain - that controls so many different functions of the mind. Additionally, she does an outstanding effort of explaining brain functioning and how it is disordered by the use of alcohol during pregnancy. Her comparison of the brain's executive functioning with the secretary of the brains true intellect or the brains Chief Executive Officer (C.E.O.) to complete the analogy is inspired and demonstrates her core understanding of ND-PAE. I think I will use that in my explanations to patients and their parents about how FASD affects the brains functioning. Having been a C.E.O. and having had a secretary, I loved the analogy of the general intellect (C.E.O. of the brain) and the executive functions (secretary of the brain) work together to get the job done. The intellect does higher level problem solving, critical thinking, abstract reasoning, judging, understanding consequences, and predicting outcomes and the executive functioning does the filing, organizing the office, scheduling appointments, sorting the mail, and other tasks that make the brain function easier. Unfortunately, in many of the patients I see both the general intellect and the executive functions of the brain is challenged. She also does a superb job of explaining sensory processing, coordination, and motor issues in people with ND-PAE by explaining older children, adolescents and adults may cope by avoiding or over-reacting in situations or environments that provoke their heightened responsiveness to stimuli, and how 'high maintenance' these individuals can be.

Dr. Susan Rich has written a complete book as there is even an appendix of parenting a child with ND-PAE. Essentially, she suggests meeting the child where they are instead of parents trying to fit them into what they think a child should be and that being strongly connected to a young child is extremely helpful to the child, but the explanation if far more detailed and helpful than these cursory remarks. Of course, she points out that children with ND-PAE are often times in environments or family contexts that do not lend themselves to warm, loving secure contacts. Her admonition to do the ABCs (Always Breathe Calmly) in these contexts is a good piece of advice, but as I learned from The Pocket Book of Mindfulness: Mindfulness isn't difficult, we just need to remember to do it - Sharon Salzberg. She also recommends using the principles of structure, supervise, schedule, and simply as a guide on how to raise children with ND-PAE. She also suggests floor-time with toddlers and as the child develops to start teaching them social skills and activities of daily living. This section is choked full of tips to engage in positive parenting and it is her contention that such strategies will a long way to secure the child a healthy future. Lastly, Appendix D is a resource for Professionals, Families and People Living with ND-PAE.

Dr. Rich is very liberal with her recognition of her colleagues who have done work to try to stem this silent epidemic often at the root cause of so much unemployment, crime, incarceration, victimization and perpetration of various forms of violence, child abuse, special education, foster care, disability, homelessness, drug and alcohol abuse, and mental illness. All through the book are model programs she has come across during her crusade to stem the tide of ND-PAE. When I was working with Dr. Satcher, 16th Surgeon General of the United States on his Children's Conference in 2000, he wisely suggested that, if the Nation wanted to do something about Public Health, it should address the problems of children in special education, foster care, and corrections. The Silent Epidemic advises everyone in the Nation about how to accomplish that goal. When I was working with Dr. Julius Richmond, President Johnson and Carter's Surgeon General and one of the architects of Head Start in the US, he taught me that in order to institutionalize something you needed a good science base, a mechanism to get the intervention accomplished, and the political will. Dr. Rich has done a venerable, exemplary job of providing the necessary science and she has given us examples and tools to get the prevention of FASD accomplished. My only hope it that the book will stimulate the political will to get the job done. One thing for sure, I will be recommending that the 4/10 patients and their families I see at Jackson Park Hospital's Family Medicine Clinic read this book as it will be immeasurably helpful to them in understanding what I have been trying to convey in the limited time I have to spend with them.

Dr. Carl C. Bell, D.L.F.A.P.A., F.A.C. Psych
Former President/C.E.O. – Community Mental Health Council, Inc. (ret.)
Former Director of the Institute of Juvenile Research (Birthplace of Child Psychiatry) (ret)
Professor of Psychiatry and Public Health – University of Illinois at Chicago (ret.)

Acknowledgments:

I would like to thank so many people for sharing their knowledge, passion, and commitment to this preventable epidemic: my mentors, past and present; my dear friends and sisters in spirit (you know who you are); patients and families who have taught me more than any book or professor; my peers, colleagues, and professional friends who have endured years of my relentless bantering and heckling to challenge the world view of children affected by prenatal alcohol exposure. I especially want to thank my children and family who love me despite years of listening to my public health messages; enduring my travels to death row, conferences, and speaking engagements; and being embarrassed in front of friends as I spoke candidly about my work.

Introduction

Imagine women who unknowingly drink a neurotoxin before finding out they are pregnant. The chemical is unavoidable because it is pervasive in drinking water yet frequently causes brain damage leading to learning disabilities, attention problems, hyperactivity, impulsivity, emotional outbursts, and even intellectual disability in 10-15% of affected children. Some women are cautious, educated, or intuitive enough to avoid it. They have the means to purchase filtered water lacking the harmful substance. Planning their pregnancies carefully, they use contraception and other means to avoid pregnancy until they are able to drink safe water. Other women unintentionally expose their babies during unplanned pregnancies, are oblivious to the chemical's presence, drink without understanding the risks or simply lack access to safe water. Unaware of the brain damage to their child, ambivalent, or otherwise unable to make a change, they drink the water despite risks to the unborn. The society is our own and the chemical is alcohol – our social drug of choice. The phenomenon, known as Neurodevelopmental Disorder associated with Prenatal Alcohol Exposure (ND-PAE), is a prevalent yet preventable cause of subtle to severe brain damage, which affects 1 in 20 American children.

Like many other people whose lives have been transformed by this epidemic, I first learned about prenatal alcohol-induced developmental disability from *The Broken Cord* by Michael Dorris. In April 1993, Dorris's provocative story about his adoptive Native American son with Fetal Alcohol Syndrome prompted me to leave a promising career in pharmaceutical research to attend public health school at the University of North Carolina. I soon realized that medical professionals – especially physicians – are as uneducated and as intellectually deaf as the lay public to the problem. Their use of the term 'funny looking kid' (or FLK) on charts of children with dysmorphic features due to this condition led me to endure the chronic pain of medical school at UNC and a 3-year residency in psychiatry at Georgetown plus a 2-year fellowship focusing on children and adolescents at Children's National Medical Center. I have since developed a successful approach to psychiatric management of these children using a neurodevelopmental treatment model.

I am a clinician working with children, adolescents and adults, not a researcher. My perspectives stem from a career encompassing pharmaceutical research, public health and medicine. Throughout the book, I share personal and professional experiences contributing to insights that prenatal alcohol exposure has caused a hidden epidemic with complex social consequences. As a board certified psychiatrist, I have encountered ambivalence among community leaders and policy makers, disbelief and misdiagnosis by physicians, struggles of parents and caregivers, and catastrophic consequences of unrecognized cases on death row. The names of real life patients and their families have been changed to protect confidentiality.

The Silent Epidemic is not about a contagious disease in the usual sense of an illness like small pox, Ebola, or HIV that is transmitted between people or communities. Instead it is about a prevalent, endemic, and commonly unrecognized preventable condition known as Fetal Alcohol Spectrum Disorder, or FASD. Fetal Alcohol Syndrome (FAS) is at the severe end of this spectrum of alcohol-induced birth defects and is represented by the most easily recognized cases. Yet, impairments in neurocognitive, adaptive functioning, and self-regulation exist to a greater or lesser degree in all individuals with FASD. This range of alcohol-related neurodevelopmental issues is listed in the psychiatrist's guidebook, the *Diagnostic and Statistical Manual fifth Edition* (DSM-5), as Neurodevelopmental Disorder associated with Prenatal Alcohol Exposure (ND-PAE). For simplification of terms, I will use ND-PAE throughout the book to reference this condition except when necessary for clarification of terms.

This book is limited to discussions about effects of prenatal alcohol exposure and is not an encyclopedia of all reproductive consequences of alcohol. While there is mounting evidence that paternal alcohol exposure changes one's susceptibility (epigenetic effects) to alcohol use disorders and possibly other conditions, the effects of the father drinking is beyond the scope of this book. In the least, fathers' involvement in helping mothers achieve sobriety by their own healthy lifestyles is as implicit in prevention efforts as supporting reproductive age women in their recovery process before and during pregnancy. Other books, such as *Drink: The Intimate Relationship between Women and Alcohol* by Ann Dowsett Johnston, provide a poignant look into the culture of drinking among women that, while related, is outside the realm of this call to action. *The Silent Epidemic* is not meant to solve the social problems so enmeshed in our collective conscience that we have become tolerant and complacent to the pervasive consequences – instead it is merely a wakeup call about prenatal alcohol-induced brain damage for the betterment of humanity.

Much of my perspectives about prenatal alcohol's hidden effects on our society are mirrored in *Brave New World,* published in 1932 – just a year before the repeal of Prohibition. The shocking truth is that the author, Aldous Huxley, knew the dirty little secret about alcohol neurotoxicity all the way back then; and so did many other educated men of his time. Capitalizing on knowledge of alcohol's harmful effects on the developing brain, Huxley, the English philosopher and French professor, envisioned a designer society by exposing certain classes to alcohol before birth. Huxley's scientifically bioengineered social structure has evolved naturally in our modern culture of self-medication, intoxication, ignorance and sexual bliss. There could be no better imagined system to stratify society than a beverage as ubiquitous and seemingly harmless as alcohol. It is not surprising that we Americans have created our own social caste system in populations where intoxication is socially condoned and half of all pregnancies are unintended.

Huxley's miscalculation was the benefits of an emotionally cocooning home with nurturing, consistent parents who model self-regulation and stability. The biologically-engineered progeny of *Brave New World* were raised in sterile nurseries with auditory programming while they slept and Pavlovian conditioning during the school day. He assumed that group raised infants would be better suited to a caste-based system than children growing up in chaotic, fractured home environments. We now recognize the neurodevelopmental harm of group rearing systems (orphanages) for any child. An understanding of the intrinsic, "hard-wiring" issues caused by prenatal alcohol exposure clarifies the "nature versus nurture" debate to explain many social problems. The neuronal wiring (nature) for children with ND-PAE leaves them at further risk for attachment issues, especially when their caregiving environment (nurture) is less than optimal. Thus, the collision of two perfect storms (ND-PAE and childhood abuse) remains an invisible yet well-defined pathway to violent antisocial tendencies. When children with ND-PAE are raised from birth in emotionally supportive, stable families with the proper knowledge, education and supports to care for a neurodevelopmentally-challenged child, the prognosis is much better than those suffering abuse, neglect, institutionalization or other early trauma.

Tragically, our indiscriminate use of alcohol during reproductive years has led to this costly yet silent epidemic through varying degrees of impairment in some exposed offspring. Numerous books for the lay public have been written describing the history, impact on individuals, and effects on society, yet social drinkers are no more informed of the risk to an unintended pregnancy than women were in the early 1960s who were prescribed thalidomide to prevent morning sickness. Why is it that books like *Drink, Message in a Bottle,* or *Damaged Angels* collect dust on shelves while more sensational books like *Take Back your Pregnancy* receive wide readership and press coverage? Society prefers to overlook the pain and suffering of lives lost within the glass of its beloved aphrodisiac. Our medical profession, public health community, and private sector can do more to prevent the unthinkable. Given that humans enjoy alcohol and sex and that over half of U.S. pregnancies are unintended —mistimed or unplanned—the time has come for physicians to encourage their sexually active, childbearing age, alcohol consuming patients to use reliable contraception. In keeping with thalidomide and Retin A policy, sensible and proactive birth control should be recommended for alcohol users. [These drugs, unlike alcohol, are classified Category X by the FDA and are not to be prescribed if there is any chance of pregnancy.]

The public deserves an understanding of the need for radical transformation – on the order of eliminating pesticides from the food chain, removing lead from paint and gasoline, and reducing smoking in public places. These are all secondary exposures to noxious chemicals that cause harm to individuals. As a society, we all agree that the risks of such exposures outweighs any benefit to the individual consumer and that protection of the public exceeds the rights of the individual. The alcohol industry, the government, and medical professionals have a duty to warn the public that alcohol causes faulty wiring in the brains of our children. Informed use of alcohol includes educating consumers about the harm to the unborn, even before we know they exist. Simply put, avoid pregnancy if using alcohol and avoid alcohol if pregnant or planning a pregnancy. The public health and mental health benefits of promoting contraceptives for alcohol consumers is synonymous with raising the legal drinking age and the effort to reduce driving under the influence. On this issue, political and religious groups should agree that the ethical right for each child to have his or her God-given healthy brain to live a morally sound, productive life outweighs any concern about promoting contraception for alcohol consumers.

The purpose of this book is threefold: 1) to highlight the magnitude of this preventable condition; 2) to enlighten physicians, caregivers, child welfare advocates, and professionals that there is hope for these children; and, 3) to inform childbearing age alcohol consumers to make educated choices. Part 1 sheds light on cases of individuals with ND-PAE from the backdrop of an affluent, suburban community to a multi-state forensic psychiatry practice. Historical anecdotes and policy references are mentioned, along with my clinical perspective and approach depending on the context of the case. Appendix I provides specific diagnostic considerations and treatment recommendations – particularly relevant for parents, caregivers, medical providers, and community professionals. Methods for identifying, intervening, and treating this condition are emphasized. While mis-wired brain circuitry will not be able to be reconstructed altogether, neuronal restructuring during the first three years of life and during adolescence can be improved by an emotionally calm, safe environment with a nurturing primary caregiver and a community of people who understand their strengths and vulnerabilities. Appendix II guides readers through a listing of national and international resources that will be updated by a website providing useful information for physicians, families and caregivers. These are dedicated to help support families and professionals.

The Silent Epidemic is not to be a stand-alone encyclopedia of ND-PAE. Instead, I am hopeful that it will inspire, motivate, and inform readers to become agents of change in this epidemic. Authors who are parents, caregivers, researchers, clinicians and affected individuals have written numerous other informational, touching and inspiring books about this topic. The book's purpose is not a war cry for prohibition or policies to promote drastic reductions in alcohol consumption. It is a hearing aid to overcome deafness to public health warnings about the leading known preventable cause of birth defects and neurodevelopmental issues in the Western world: prenatal alcohol exposure. We all need to recognize that prevention includes promoting awareness about preconception health, contraceptive use (i.e., pregnancy prevention) for alcohol consumers, as well as avoiding alcohol if pregnant or planning a pregnancy.

Tiring of the painstaking work reviewing difficult case histories of individuals who fell to death row through the gaping holes in our broken mental health system, I have redirected my efforts back upstream to public health. To that end, in 2014, I founded a nonprofit – 7th Generation Foundation, Inc. and started a prevention blog[1] with a 25-year-old psychology graduate student from Germany, Daniela Mielke. 7th Generation Foundation aims to promote awareness about ND-PAE, prevent the condition through preconception health and safe sex for alcohol consumers, and to develop safe, affordable housing and vocational supports for affected individuals. The proceeds from this book will (in part) provide the financial means to accomplish these efforts.

---

[1] www.bettersafethansorryproject.wordpress.com

At the time of writing of this book, I am hopeful that our society is on the verge of recognizing the silent epidemic happening to children, adolescents and young adults with ND-PAE. In the last two months, Kathleen Mitchell of the National Organization on Fetal Alcohol Syndrome has been on several national television shows after being interviewed for an article in the *Washington Post* about her 43-year-old biological daughter who has Fetal Alcohol Syndrome (FAS).[2] Last month, the Centers for Disease Control and Prevention issued a statement that alcohol consumers should use reliable contraception to prevent alcohol-related embryopathy (birth defects and neurodevelopmental problems associated with exposure to alcohol prior to pregnancy recognition). This year for Fetal Alcohol Spectrum Disorder (FASD) Awareness Day, I am planning to have a "Millions who Recognize FASD March on Washington," similar to the Million Man March and the Million Moms March.

Hopefully, sharing aspects of my personal and professional journey will create lasting societal change to protect our most valuable human potential – our children and the unborn yet to be … for generations to come.

In the spirit of healthier futures,

Susan D. Rich, MD, MPH, DFAPA
Child/Adolescent and Adult Psychiatrist
Distinguished Fellow, American Psychiatric Association

---

[2] Fleming AR. This mother drank while pregnant. Here's what her daughter's like at 43. *Washington Post*; January 18, 2016. https://www.washingtonpost.com/national/health-science/this-mother-drank-while-pregnant-heres-what-her-daughters-like-at-43/2016/01/15/32ff5238-9a08-11e5-b499-76cbec161973_story.html

Part I

Illuminating the Problem

WHAT IS NEURODEVELOPMENTAL DISORDER ASSOCIATED WITH PRENATAL ALCOHOL EXPOSURE?

The other specified neurodevelopmental disorder category is used in situations in which the clinician chooses to communicate the specific reason that the presentation does not meet the criteria for any specific neurodevelopmental disorder. This is done by recording "other specified neurodevelopmental disorder" followed by the specific reason (e.g., "neurodevelopmental disorder associated with prenatal alcohol exposure").[3]

The Diagnostic and Statistical Manual, Fifth Edition lists Neurodevelopmental Disorder associated with Prenatal Alcohol Exposure (ND-PAE) under Specified Other Neurodevelopmental Disorder (ICD-9 Code 315.8). ND-PAE is a diagnosis to explain the complex neuropsychiatric, cognitive, behavioral, social, language, communication and other multi-sensory deficits associated with maternal alcohol use during pregnancy. In Section 3 of DSM-5, criteria have been proposed for ND-PAE diagnosis (in Appendices A-2 and A-3). While these criteria have not been evaluated for sensitivity and specificity of diagnosis in large numbers of populations, more than 40 years of research has contributed to understanding the wide array of brain damage associated with prenatal alcohol exposure.

---

[3] *Diagnostic and Statistical Manual of Mental Disorders*, 5th Edition (DSM-5). American Psychiatric Association, 2013, p. 86.

This condition has been described in research and the lay press as Fetal Alcohol Spectrum Disorder (FASD), an umbrella term for a variety of conditions ranging from Fetal Alcohol Syndrome (FAS) to Fetal Alcohol Effects (FAE) to Alcohol-related Neurodevelopmental Disorder (ARND). The term ND-PAE was first published in May 2013 in the Diagnostic and Statistical Manual of Mental Disorders, 5th Edition (DSM-5) yet has existed as long as mankind has imbibed alcohol. Since alcohol disproportionately affects the developing brain during pregnancy, babies are often unknowingly exposed before a woman even knows she is pregnant.

Their symptoms from birth mimic a myriad of neuropsychiatric conditions, often confusing or overwhelming the most seasoned physician or mental health professional. A child with ND-PAE is born with brain wiring differences ("intrinsic vulnerabilities"). If they grow up in an environment of tolerance for their behaviors with adequate supports, early interventions, appropriate structure, and unconditional love and nurturing by parents who have realistic expectations, this child will have a much better chance to accept and compensate for their neurodevelopmental challenges. In contrast, a child with ND-PAE who is not diagnosed may feel different from peers, struggle with a harsh, abusive, or otherwise traumatic childhood with a neglectful, ambivalent, or punitive parenting style, and lack school supports to succeed in transitioning to adulthood.

Early screening and intervention services help develop a child's cognitive potential, improve their social communication skills, and overcome their sensory and motor issues. However, neurodevelopmental effects of prenatal alcohol exposure are lifelong, permanent, and disabling. Adaptive functions and executive functions are frequently affected as are mood regulation and autonomic arousal (i.e., "fight or flight"), social communication abilities, and sensory or motor functions. In turn, ND-PAE can lead to psychological and psychiatric factors that create intrinsic vulnerability to academic challenges as well as externalizing, defensive, oppositional, defiant, maladaptive, and acting out behaviors. Parents and school staff having realistic expectations on the child's ability to adapt to the world around them and to accomplish age-appropriate tasks reduces the child's frustration and anxiety, improving their mood regulation by environmental supports.

Commonly, children with prenatal alcohol exposure have relatively average to below average intellect, reading and math disorders, attention deficits, language and social problems. Social anxiety can evolve into either agoraphobia (being afraid to go out of the house, fear of crowds or large spaces, etc.) or paranoia (being afraid someone was out to get him or harm him in some way). Many adolescents and young adults with ND-PAE in my practice suffer from heightened stress response that can evolve without treatment into severe suspiciousness, paranoia, and delusional thinking that mimics psychosis.

Extrinsic factors, such as witnessing or experiencing abuse, neglect, loss, and other trauma has been shown to further impact neurodevelopment. These children may develop relational problems, such as reactive attachment disorder, oppositional defiant disorder, and conduct disorder as well as personality disorders (antisocial personality disorder, borderline personality disorder, etc.). The chapter on *Nature versus Nurture* and Appendix C: *Parenting a Child with ND-PAE* provide greater detail on the impact of environment on a child's prognosis.

One of the main reasons that children are not identified with ND-PAE is because the mothers may not admit use during early stages of pregnancy or their physician may only ask, "You didn't use alcohol during pregnancy, did you?" A better, less threatening approach would begin with questions about the pregnancy intention status (mistimed or unplanned), the point at which they found out they are pregnant, and their lifestyle behaviors before, during and after pregnancy. Non-judgmental discussions about effects on brain development prior to pregnancy recognition and the lack of public understanding about this little known fact can help bridge the mother's understanding without making her defensive.

Traditional psychiatric evaluations include obtaining a thorough history of symptoms and assessing observable behaviors in a clinical setting that can sometimes be measured through questionnaires completed by parents, caregivers, teachers, and self-reports. For patients presenting with a complex array of issues, ND-PAE should be at the top of the differential. Forms in Appendices A-1 to A-4 can be downloaded and used clinical settings for screenings of patients. Astute clinicians are adept at knowing when to use standardized questionnaires, refer for neurological, neuropsychological, speech/language, occupational, and/or physical therapy evaluations. Assessments of other issues such as depression and anxiety disorders can include evaluation of observed behaviors but often involves careful interviewing techniques to elicit underlying feelings and emotions.

Babies who are poorly regulated in their sleep/wake cycle, are difficult to soothe or comfort, or have hyper- or hyposensitivity to light, sound, taste, temperature, textures or other experiences may have underlying neurodevelopmental issues, such as autism, ADHD, or ND-PAE. It is common for children with undiagnosed ND-PAE to have been expelled from preschool more than once, to be labelled with "behavioral issues" in elementary school, to get into fights with peers and have oppositional behaviors over homework or chores with parents in middle school, and to have academic failure, juvenile delinquency, or running away in high school. In my experience, these and more severe consequences are avoidable by accurate diagnosis, appropriate treatment and interventions, and proper transitional supports into adulthood. More importantly, by helping parents and caregivers understand the brain-based underpinnings of the child's behaviors, they can adjust their expectations, parenting strategies, and advocacy efforts for their child. While individuals with ND-PAE have a number of challenges, they also are blessed with strengths and interests which can develop into vocational skills toward a productive work life.

The first chapter in this book raises the concern about undiagnosed and untreated ND-PAE with a history of trauma as a contributor to catastrophic outcomes, as exemplified in the case of Adam Lanza ...

Chapter 1

I Am [not] Adam Lanza's Psychiatrist

Author's disclaimer: The information presented in this chapter was gathered from public information published about the event, the troubled young man who has been referred to as "the shooter," and his mother—a lovely woman and caring mother who needed help for herself and her son. It presents a poignant example of a possibly unrecognized and untreated individual with neurodevelopmental disorder associated with prenatal alcohol exposure. Although provocative and perhaps offensive to some, my inclusion of this as a first chapter provides a call to action for understanding and treatment of this preventable condition. The chapter title is a spin-off of a blog, "I am Adam Lanza's Mother," written by Liza Long in the Blue Review.[4]

The morning of December 14, 2012, I walked my six-year-old kindergartener and nine-year-old third grader into their Spanish enrichment classes in our quiet community in suburban Maryland. As that unremarkable workday began, I didn't question whether I would see my children after school. Amidst a morning of report writing and seeing patients, the unprovoked slaughter of so many innocent children and adults at Sandy Hook Elementary in Connecticut silently unfolded. Patients shared descriptions of the unthinkable tragedy they had heard on the news. Shaken and dismayed by the accounts, my heart ached for the parents of those young children and the school staff. For a

---

[4] Long L. I am Adam Lanza's Mother. *The Blue Review*; 12/15/2012. http://thebluereview.org/i-am-adam-lanzas-mother/

parent, possibly the worst nightmare is losing a child to such a cruel and insensitive act of violence. Throughout the day, I said a selfish prayer: "Dear God, please be with the families in Connecticut, and please be with my children."

Helping others through stress, fear, and turmoil in their lives limits my coming unraveled at problems that do not directly affect me, my family, or my patients. With the holidays rapidly approaching, professional and personal obligations superseded my watching media coverage about the shootings. And because my own kids were similar in age to the innocent children in Connecticut, I did not allow myself to dwell on and become even more upset about the senseless act of violence. Being the daughter of a police officer who blossomed into a child psychiatrist, I had learned early in life to compartmentalize and diffuse rather than dramatize and exaggerate catastrophe.

For several days after the deadly, heinous event, I avoided the radio, newspapers, Internet, and news to maintain peace and balance in my own life. Nonetheless, over the intervening days, I began to hear snippets from patients, friends, and neighbors. One mother told me that the young man perpetrating the shootings seemed slightly autistic. A girlfriend said he had sensory issues that led to him wearing noise-reducing headphones during the incident. My sister-in-law mentioned she heard he had social and academic issues, leading him to leave college and return home to live with his mother.

The occasional fleeting thought arose amid the media hype that he might have ND-PAE. During a brief exchange by phone with a mom in my practice, she remarked that he resembled her adopted daughter from Russia. I replied that his profile was chillingly similar to individuals with ND-PAE I had interviewed on death row. Days later, while flipping the cable channel between wrapping gifts, I gasped. The vacant eyes and gaunt face of a young man who seemed to have Fetal Alcohol Syndrome (FAS) stared back at me. The reporter identified him as Adam Lanza, the perpetrator of the mass murders at Sandy Hook Elementary. How is it that both I and my patient's mother recognized FAS in his face and history, yet hardly anyone else in the world seemed to notice?

There were many red flags in his history to indicate he was affected by prenatal alcohol exposure.

Media reports painted a picture of a boy who was a socially-awkward, introverted child who had developed into an antisocial, isolated, angry young man. Former classmates described him as "bright, but extremely shy and remote."[5] The *Huffington Post* reported that when "people approached him in the hallways, he would press himself against the wall or walk in a different direction, clutching tight to his black briefcase. He was often having crises that only his mother could defuse."[6] Some of his anxious episodes led him to run away to avoid being in class or doing schoolwork. His autistic-like behaviors, social communication, learning issues, dropping out of college, difficulty "launching" into an independent lifestyle, and rage issues were difficult even for his mother to control.[7]

Radio broadcasters noted that his mother was well known at a local bar and portrayed her as a big-time "party girl," although also was somewhat reclusive even to members of her book club. Since ND-PAE can happen as early as the third week after conception, many women expose their babies to their binge drinking before knowing they are pregnant. During the years around his birth in April 1992, women viewed use of alcohol during pregnancy very differently than when the warning label came out in 1981. Yet whether Nancy Lanza used alcohol during pregnancy likely went to her grave.

In May of 1997, my aunt had clipped and sent me a newspaper article about the hallmark findings of increased drinking in pregnancy from 1991 to 1995. Maternal alcohol use rose by 400 percent from 1991 to 1995, according to the Centers for Disease Control and Prevention.[8] With the article in hand, I left for an internship as a medical student with ABC's *Good Morning America*. After pitching the story idea to the medical producer of the *Healthy Woman,* together we produced a

---

[5] http://investigations.nbcnews.com/_news/2012/12/15/15935397-mom-of-suspected-school-shooter-first-to-die-was-avid-gun-enthusiast-friend-says

[6] http://www.huffingtonpost.com/2012/12/16/nancy-lanza-mother-of-gunman-private-home-life_n_2313027.html

[7] Ibid.

[8] *Morbidity and Mortality Weekly* 46, no. 16 (April 25, 1997): 346–350.

segment warning women to use contraception if using alcohol and having sex, and to plan their pregnancies very carefully to avoid unintentional prenatal alcohol exposure. Now, as I sat wrapping Christmas gifts fifteen years later, I saw firsthand a likely bi-product of that epidemic.

Overcoming dismay, shock, and revulsion for the massacre, I speculated about hard-wiring flaws causing his impaired mental state at the time of the senseless crime. Immediately scouring the Internet for information about the case to see if anyone else suspected that he had ND-PAE, I realized I was not alone. Others had the same idea, reporting on their personal blogs that his facial features and low-set ears appeared as in those with FAS. Based on his behaviors, physical features, self-regulation, and learning and social challenges, he was a young man who suffered from ND-PAE.

A variety of television reports mentioned that investigators conducted genetic tests to look for hereditary problems (fragile X syndrome, chromosome mutations, and other genetic disorders) linked to the young man's mental state. Doctors such as me, working with children who have complex neurodevelopmental issues, often test for and explore possible genetic and prenatal exposures such as alcohol. These organic issues can be linked to defective brain wiring associated with maladaptive, impulsive, and violent behaviors. Without a doubt, Lanza was a seriously ill young man with a chronic, undiagnosed neurodevelopmental condition. Over the years, however, doctors likely never asked his upper-class mother whether she drank when she was pregnant or prior to pregnancy recognition.

My insight into the connection between prenatal alcohol exposure and juvenile violence is mirrored in *The Fatal Link: The Connection between School Shooters and the Brain Damage from Prenatal Exposure to Alcohol*[9] by Jody Allen Crowe. Like me, Crowe's perspectives come from his personal and professional experiences, having graduated from the high school where the first mass school shooting occurred in 1966 in Grand Rapids, Minnesota, to his work as an educator on Native

---

[9] Jody Allen Crowe. *The Fatal Link: The Connection between School Shooters and the Brain Damage from Prenatal Exposure to Alcohol*. Denver: Outskirts Press, Inc., November 25, 2008).

American reservations. During the summer of 2015, Jody presented with me and three of my interns at the International Congress on the Law and Mental Health in Vienna, Austria. Sharing his insights about the Red Lake rampage that shook the nation, he was dismayed that media reports failed to acknowledge the possible effects of prenatal alcohol exposure on the shooter's mental status—despite features of FAS in the shooter's face and recognition that his mother was a severe alcoholic. He also found evidence of unrecognized ND-PAE in cases of school shooters in Wisconsin and Minnesota all the way back to forty years ago.

The catastrophic case of Adam Lanza and other school tragedies outlined in Crowe's book suggests to me the need for drastic paradigm shifts in our society's awareness about mental illness and ND-PAE, the leading cause of preventable brain damage affecting our children. Unfortunately, few mental health providers know much about ND-PAE and sometimes make matters worse by overmedication, inadequate inpatient and residential care, and improper parent guidance. While those children and adolescents who receive treatment for mental illnesses are no more likely to commit violent acts than the average adolescent, the costs of not identifying and treating ND-PAE and other neuropsychiatric disorders can be devastating for our communities, families, and children.

Adam Lanza is like countless other complex patients with multiple disorders who have received less than optimal treatment and community support. According to the American Academy of Child and Adolescent Psychiatry, around 20 percent (i.e., seven to twelve million) of American youth are affected by mental illness, and only 20 percent ever receive treatment.[10] As a whole, the field of child/adolescent psychiatry and psychology is woefully understaffed and underfunded, with the field being in the category of "physician manpower shortage area" since the 1970s. Even if every child and teen who needs treatment were to seek services, providers and appropriate hospitals would only be available in certain locations in the country.

---

[10] http://www.aacap.org/App_Themes/AACAP/docs/Advocacy/policy_resources/Children's_Mental_Health_Fact_Sheet_FINAL.pdf

In my own practice, one twelve-year-old girl with intellectual disability associated with prenatal alcohol exposure waited in the emergency room for more than ten days before being placed in a psychiatric ward not equipped to handle her neurodevelopmental issues and cognitive deficits. A nineteen-year-old patient with ND-PAE waited for three weeks in the emergency room before her parents decided to take her to a therapeutic foster home, where she is currently receiving comprehensive services for reactive attachment disorder. In speaking with the medical director of the hospital, he said it is not uncommon to have a young person wait six months for a bed, stuck in purgatory of the emergency room without receiving meaningful treatment. Others languish in prisons and jail cells, awaiting placement on involuntary commitments. Improving our society's commitment to child psychiatry would improve psychological and emotional development for the most vulnerable of our children.

A recent comprehensive report on Adam Lanza highlighted a lack of response from the educational and healthcare systems as well as untreated mental illness as primary contributors to Adam Lanza's progressively worsening condition. Several recommendations mentioned in the report:

- ✓ universal mental health screening for children (birth through age twenty-one)
- ✓ better training for educators
- ✓ better support programs for families
- ✓ better financial support for school-based therapy-related services (e.g., occupational therapy and behaviorist services)
- ✓ additional support to provide in-school services for children with 'highly specialized' needs. [11]

Had I been Adam Lanza's psychiatrist, I would have helped develop a well-rounded treatment plan as the psychiatrist who evaluated him most likely attempted to do. The comprehensive services would have included socialization activities, such as socially-engineered playgroups from an early age, boy scouts or a similar hands-on recreational activity,

---

[11] "Shooting at Sandy Hook Elementary School: Report of the Office of the Child Advocate." State of Connecticutt,. November 21, 2014.

and therapeutic social skills group training; farm animal or equine therapy; parent guidance and education; as well as involvement in healthier outlets (e.g., team or individual sports and/or recreational activities) to replace his obsession with video games. I would have recommended that his mother, and father, take him for walks and hikes in nature while camping in mature hardwood forests; exploring caverns; and spending time on farms caring for docile domesticated animals. There are often years of psychosocial interventions necessary to prevent catastrophes, such as Sandy Hook. Despite earlier missed opportunities for healing, Adam Lanza likely needed hospitalization followed closely by residential treatment during the weeks leading up to the massacre.

So often even in residential treatment centers, children and adolescents with ND-PAE go unnoticed by staff and misdiagnosed as oppositional/defiant, garden-variety ADHD, bipolar, anxious or depressed by psychiatrists, and overmedicated on a number of highly potent antipsychotics and mood stabilizers. The goal becomes "chemical restraint" rather than understanding the adolescent's brain wiring issues. Often, professionals fear that a diagnosis of ND-PAE is a death sentence – stigmatizing the individual to a life in the underclass. More training is needed for medical professionals and mental health providers to understand that. The proper diagnosis allows a right to a life of purpose, dignity, and strength through understanding, treatment, and commitment on the part of the community. Educating primary care physicians about ND-PAE who may see these children in their practices but may not understand the impact of a mother's drinking behaviors prior to pregnancy recognition would help identify more of these children early enough to benefit from a coordinated systems of care approach. These policies for identification of children must be incorporated into better access to outpatient mental health and substance abuse services for affected individuals.

In my view, even if we identified every single one of the "1 in 20" children with ND-PAE in our country, there would not be enough programs and services to meet their complex needs. Specialized school programs for children with ND-PAE are needed in this country to help teach them based on their learning styles, to support their challenges and nurture their strengths, to teach emotional regulation and instill a moral compass through character development, and to provide an avenue for their success and productivity (i.e., apprenticeship programs, vocational studies, recreation and hobbies, life skills training, social skills classes, speech and language programs). We must embrace families with children affected by our most favored beverage, helping them raise their children with the proper supports, education, vocational assistance, housing, and health care benefits. Identifying, treating, and preventing this condition will help move our society into a new age of reduced juvenile delinquency, violence, and other secondary outcomes of undiagnosed ND-PAE.

Over twenty-two years ago, I entered the field of public health with an understanding of the long term consequences of undiagnosed and untreated prenatal alcohol exposure. At that time, the research community had joined with parent advocacy organizations to promote awareness and develop resources for affected families. Researchers and caregivers described a range of physical, developmental, emotional and behavioral effects associated with prenatal alcohol exposure. Dr. Anne Streissguth, a pioneer in diagnosis and treatment of affected individuals at the University of Washington at Seattle described the continuum of effects as an "iceberg" phenomenon. The "tip" above the water line is made up of children with dysmorphic features evident to the naked eye. The majority of affected individuals have neurodevelopmental deficits, less visible to the naked eye, and less pronounced physical features. A concise explanation of this phenomenon comes from founder of the FAS Family Resource Institute, Jocie DeVries, and founding editor of *Iceberg* in Washington State, Dale Leuthold, whose pioneering advocacy have provided important resources for families across the country.[12]

---

[12] DeVries J with McCann D. DSM-V: Making the Case for an FASD Behavioral Phenotype. *Iceberg* Newsletter, FAS Information Service, February 12, 2010. http://www.fasiceberg.org/newsletters/vol20num1_mar2010.htm#psychiatry

The Silent Epidemic: A Child Psychiatrist's Journey beyond Death Row

Figure 1. The "Iceberg" Phenomenon – Most cases are hidden below water level: among the socially disenfranchised, vulnerable, complex, and misunderstood clients frequenting homeless shelters, psychiatric and criminal justice systems.

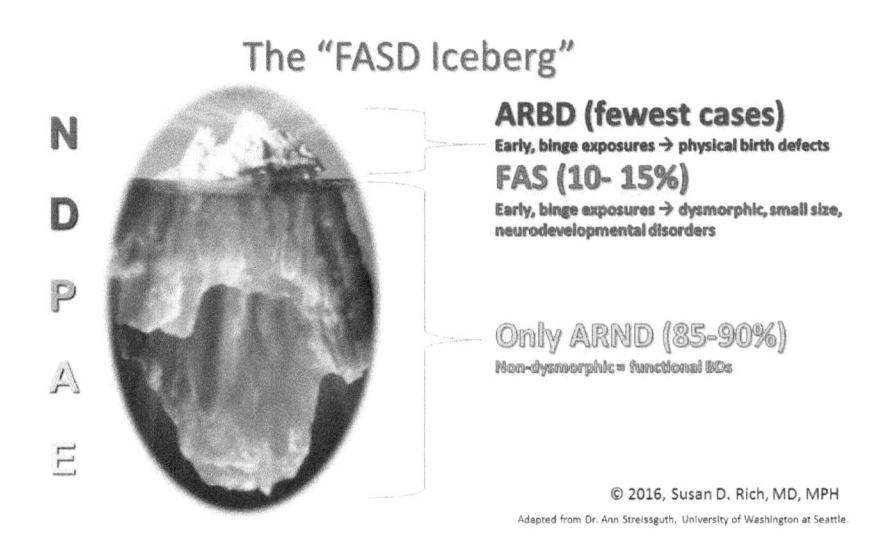

The "FASD Iceberg"

N D P A E

ARBD (fewest cases)
Early, binge exposures → physical birth defects

FAS (10- 15%)
Early, binge exposures → dysmorphic, small size, neurodevelopmental disorders

Only ARND (85-90%)
Non-dysmorphic = functional BDs

© 2016, Susan D. Rich, MD, MPH
Adapted from Dr. Ann Streissguth, University of Washington at Seattle.

Ann Streissguth, Ph.D., and her colleagues had been studying the effects of prenatal alcohol exposure since 1973 and were well aware of what was lurking beneath the waterline ... Dr. Streissguth became famous in the research world for her statement that FAS was only the tip of the iceberg.

The [iceberg] symbolizes a concept that applies both to fetal alcohol syndrome/fetal alcohol effect (FAS/FAE), and to the powerful social, political, and personal ramifications that are associated with this disease. A small part of the problem is visible – but the bulk of the problem is obscured and hidden.

The problem is enormous and complex ... Like an iceberg, the visible portion of the problem gives us warning of hidden danger. We are not sure of the extent of that danger. We don't know how big it is. We are sure the danger is real, but we

> *are in the process of learning how to best use our resources and energies to fight it.*[13]

As a child psychiatrist, I believe Adam Lanza's pattern of physical, neurodevelopmental, and emotional characteristics were hauntingly similar to individuals with features below the waterline of the iceberg, having parents who were ill-equipped and distraught about how to help him. Their reluctance to insist that he receive treatment caused him to suffer in silence behind closed doors at home rather than face the social repercussions of psychiatric care. What responsibility do parents have to protect others when a psychiatrist or other mental health professional deems them in need of services but the parents fail to follow through? Would his parents have let him go without treatment if he had juvenile diabetes or high blood pressure? What about a broken bone or a concussion?

Our society must shed the veil of stigma associated with brain-based neurodevelopmental conditions, much like we have promoted policies for inclusion of individuals with physical limitations (i.e., blindness, paraplegia, loss of limbs, etc.). We must ensure funding for improved mental health programming for children and adolescents, including intervening if necessary even when parents are reluctant. In the absence of early, accurate diagnosis, a majority of unfortunate individuals with ND-PAE end up in catastrophic circumstances similar to Adam Lanza. Proper diagnosis, treatment, and support can improve their prognosis by giving them the services necessary to live relatively productive lives. Without diagnosis and treatment, it is the march of the penguins off the iceberg to death row for far too many individuals.

No matter the precipitants, the act of merciless violence in Newton, CT set fire to my passion to inform the public of the need to prevent ND-PAE. Just as I felt compelled to take on this cause in 1993, a renewed call to action 20 years later came in the form of the unspeakable tragedy at Sandy Hook. It has been my opinion for more than two decades that such senseless acts of rage and disinhibited violence are worthy of our

---

[13] Streissguth AP. Birth and Life of an Iceberg. *Iceberg* Newsletter. FAS Information Service; February 12, 2010. http://www.fasiceberg.org/newsletters/vol20num1_mar2010.htm#psychiatry

society acknowledging root causes of violent and destructive acts to better protect our children and communities. This work has taken me to maximum security prisons in Virginia, Pennsylvania, Georgia and Tennessee. Like other esteemed forensic colleagues, I have examined and provided expert testimony to help judges, prosecutors and public defenders understand the brain damage associated with prenatal alcohol exposure, particularly as a mitigating factor in capital murder cases.

Fortunately, my journey did not begin or end on death row. I have had the privilege to venture there temporarily as a witness to humanity's failure to solve its social disparities in an ethically-responsible way. Hopefully, insights into lives of individuals with this disorder now behind bars will create lasting societal change to protect all our children – even the unborn yet to be.

Chapter 2

Behind Bars

An enquiry into the cause of the late increase in robbers ... with proposals for remedying this growing evil ... What must become an infant who is conceived in Gin ... with the poisonous distillations of which it is nourished, both in the Womb and at the Breast?

— Henry Felding, London: A. Millar; 1751.

    The click-clack of my high heels echoed off the freshly waxed tile floor in cadence with the guard's footsteps like horse hooves on cobblestone streets. Steel doors sealed off endless hallways, interrupting our solitary march to death row. Imprisoned by a maze of maximum security, there was no turning back. With the last door clanging behind, I was captured, entombed with society's 'untouchables.' I shuddered softly as the tall, well-built guard escorted me into a small, sterile exam room. In the center of the room, Jack Ford – a short, stocky man – unshaven, wearing a tee shirt and lounge pants – sat in handcuffs shackled to a table. His rough exterior spoke to years of hard labor, relegated to the working poor of Huxley's delta class. His white crew cut framed a 50'ish year old face with unusual features – short, upturned nose; thin, flattened upper lip; and wide set, slightly slanted eyes. His astute public defender recognized the possibility of prenatally-induced brain damage in the defendants face and sought out my help. Such tell-tell signs of Fetal Alcohol Syndrome (FAS) are rarely documented in adults because adolescent development minimizes the "dysmorphic" (odd-looking) appearance.

I was asked to provide a professional opinion about Jack Ford because of my expertise in evaluating neurodevelopmental disorder associated with prenatal alcohol exposure (ND-PAE). In a way, my job in capital murder cases where a person's life is on the line is to seek out information, much like a bloodhound searching for clues of living victims in a natural disaster. I search for evidence of brain damage caused by a mother drinking during pregnancy – from affidavits of family members or spouses, from magnetic resonance imaging showing characteristic brain damage, from school records indicating learning disabilities or attention/hyperactivity issues, and other related clues.

In my forensic role, I am seldom able to interview the mothers, only reading about their own histories of drinking habitually to numb the pain of child abuse, incest, or domestic violence. A majority of the time, pregnant mothers are not intentionally using alcohol to harm their babies but instead to cope with the problems they feel helpless and hopeless to overcome. My heart breaks as I read about the years of physical, emotional, and verbal abuse at the hands of their husbands or fathers, using alcohol to self-medicate the pain and psychological wounds they endured. Often, many of the mothers have been exposed prenatally to alcohol themselves and lack the life skills and coping strategies to pull themselves out of abusive relationships.

Jack's mother was a battered woman who, by affidavit of several family members, drank heavily throughout her childbearing years. No one intervened to stop her drinking, provide support for her sobriety, or encourage her to avoid pregnancy until she could get sober. So, there she sat sixty years ago on the outskirts of a rural coal mining town, year after year sinking into a deeper hole of misery for her and her children. It is no wonder that some of the children ended up in special education classes and Jack flunking out of school.

Driving to the prison that morning from a roadside motel in a sleepy mountain hollow brought flashbacks to "Deliverance." It felt surreal to be venturing to death row while breathing in the beautiful, rural countryside. The well-kept grounds and neatly groomed flower beds of the maximum security facility belied the pain and anguish of souls held in purgatory. The institution could as easily house a vocational school or homeless shelter as much as a warehouse for society's castaways. Unlike its lifelong residents, I ventured into that isolated and forgotten catacomb knowing I would be going home at the end of the day.

What brought me to death row was a link between trauma and prenatal brain damage that we know all too well from criminal justice research. It is the combination of witnessing or experiencing abuse and "neurodevelopmental" issues that predispose a person to violent and persistent antisocial (criminal) behavior. Several decades of elegant, costly, large-scale research using population based epidemiology and criminal mapping methods have proven this little known fact. Since the 1950s, Wolfgang, Lewis, Moffitt, Raine, and many other criminologists have independently described traits predisposing individuals to criminal behavior. While in graduate school during the early 1990s, I became interested in this largely unrecognized link - an association between antisocial behaviors and the interplay between prenatal alcohol-induced brain damage and histories of childhood trauma.

Developmental and learning disabilities have long been known to be related to trouble with the law, homelessness, and employment problems. Individuals who have neurodevelopmental problems and witness or experience abuse are much more likely to perpetrate violent and persistent antisocial behavior. Largely overlooked from this large body of research is the red polka dot elephant in the room. Children from alcoholic and substance abusing homes frequently have prenatal alcohol exposure and witness or experience abuse. I proposed this link in 1994 in my master's thesis, *A Link between Fetal Alcohol Effects and Juvenile Crime*[14] (Figure 2). Therefore, we would eliminate a significant pre-determinant to crime and violence by reducing prenatal alcohol exposure, not just child abuse and neglect.

---

[14] Rich SD. A Previously Unexamined Source of Delinquency: Fetal Alcohol Effects - An Emerging Paradigm. The University of North Carolina, School of Public Health, Master of Public Health thesis, 1993-94.

Figure 2. Conceptual Link between ND-PAE, Childhood Trauma, and Antisocial Behavior

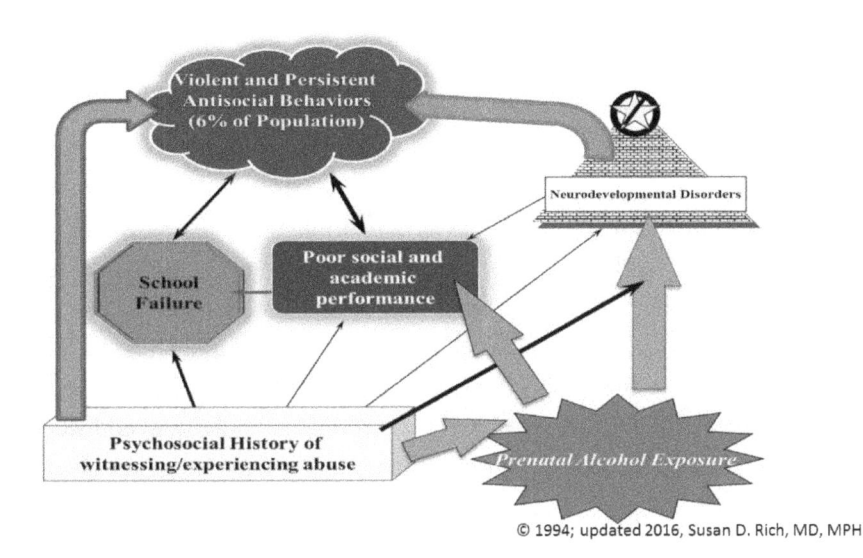

© 1994; updated 2016, Susan D. Rich, MD, MPH

In short, children with ND-PAE raised in stressful, physically violent homes see the world as hostile and unsafe. In many cases, they begin life with mild to moderate brain damage, have psychiatric illnesses from the psychological trauma they grew up with, and fight their way through life. Underlying learning disabilities, illiteracy, and frustration with higher level classes by middle and high school lead to academic failure. Because the school systems are set up today to teach academic skills rather than infusing the curriculum with vocational and life skills, adolescents do not learn any reasonable job skills for gainful employment. The combination of these factors leads to increased tendency toward poor school performance, un-employability, and criminal behavior.

This bio-psychosocial model of criminal behavior was published in *Ghosts from the Nursery* by Robin Karr-Morse and Meredith S. Wiley in 1997. Just before entering medical school in 1996, I met Ms. Karr-Morse at NATO-sponsored conference on Rhodes Island, Greece. They devoted one section in their chapter on prenatal exposure to drugs and malnutrition to prenatal alcohol exposure. They mention at least one study from the early 1990s indicating as little as 0.5 ounce of alcohol per day causes clinically significant effects on mental performance linked to a propensity to violence.

A nearly 60-year-old example of the failure of our fractured system of care sat in front of me – Jack Ford. A number of family members reported that Mrs. Ford drank heavily throughout her pregnancy with Jack. Because of his neurodevelopmental issues caused by prenatal alcohol exposure, Jack had difficulty learning, was bullied for being "slow" by classmates, and seldom found solitude in the schoolyard. His older sister said that kids would make fun of Jack, calling him a "retard" but that he never got into trouble at school. She stated that she and Jack wanted to go to school and wanted to learn but that they had a lot of trouble with the work. Often, when he did attend school, he would sleep through most of his classes, especially during the winter months. From an early age (8-9 years old), as one of several children, he had to dig coal out of a slate dump to keep the fire going at night. He recalled that his dad who was 6'1" and weighed 220 pounds would strip him naked and beat him with a hose if the fire went out and the family got cold. Sadly, no one ever checked in on Jack or provided adequate support for his disabilities. One of Jack's younger brothers recalled that Jack dropped out of high school when he was 14 or 15, and when asked whether anyone came to check on Jack about truancy, he remarked – "they didn't care if we came to school or not, nobody came to find [Jack]."

Knowledge about the links between academic failure and criminal behavior, violence, and incarceration has led to downstream rather than upstream approaches by criminal justice experts. "Profiling data" for youth with reading difficulties forms the basis for policies enacted into local, state and national law. A colleague, mentor, and co-presenter,[15] the late professor Charles Dean, shared an alarming yet controversial fact. State and municipal governments, such as Alabama,[16] review standardized test scores to determine percentage of children unable to

---

[15] Rich SD and Dean CW. A Previously Unexamined Source of Delinquency: Fetal Alcohol Effects - An Emerging Paradigm. Academy of Criminal Justice Sciences Conference (Mar 96, Las Vegas) and the American Society of Criminology Conference (Nov 96, Chicago).

[16] *Arizona Republic*, 9-15-2004.

read by fourth grade to estimate numbers of jail and prison beds that will be needed when that cohort reaches young adulthood. When I have presented this obscure fact at conferences, responses vary from disgust and disbelief, to agreement and acceptance of the practice.

Literacy statistics and juvenile delinquency are well touted in the lay press. We have known for decades that children with learning issues and cognitive impairment fall through the cracks in our fractured educational and mental health system, filtering to the bottom tier of society (the underclass) and becoming more and more socially disenfranchised. Evidence shows that children who do not read by third grade often fail to catch up and are more likely to drop out of school, take drugs, or go to prison. Low literacy is strongly related to crime, with 70% of prisoners reading at or below the fourth grade level. Low literacy is strongly related to unemployment, with more than 20% of adults reading at or below the fifth grade level – far below the level needed to earn a living wage, especially in a growing technical society.

Nicholas D. Kristof lists illiteracy as a major link with juvenile crime in *Profiting from a Child's Illiteracy*.[17] He cites 85% of all juveniles who interface with the juvenile court system and 60% of all prison inmates as functionally illiterate. The Department of Justice has linked reading failure to academic failure, delinquency, violence and crime. Penal institution records show that inmates have only a 16% chance of returning to prison if they receive literacy help, as opposed to 70% who receive no help. Annual taxpayer costs are $25,000 per inmate and nearly double that amount for juvenile offenders.

Jack's trouble with literacy followed him into adulthood when he attempted to take the truck driver certification test. His wife worked with him for several hours each day to memorize the material for the test, and in the end, he took an oral version of the test. As a truck driver, he always had his wife or son ride along with him to read the maps and help him with directions. He had used drugs and alcohol throughout his adult life and had three failed marriages by his late 50s. Finally,

---

[17] Nicholas D. Kristof. *Profiting from a Child's Illiteracy, New York Times;* 12/7/12.

someone took notice after a series of impulsive events culminated in his girlfriend's death.

While awaiting a death sentence, his facial features struck an astute social worker with public defender's office, reminding her of images she'd seen of individuals with Fetal Alcohol Syndrome. She immediately began researching his birth history, school records, and other related documents to find out there was evidence of intellectual disability, attention deficit hyperactivity disorder, social skills problems, and other "red flags" for ND-PAE. Soon after, she and Jack's team of attorneys reached out to me and a neuropsychologist colleague for our assistance.

Jack's case is only one of countless unfortunate failures of our society to adequately understand and deal with individuals with ND-PAE. Their limited social skills, emotional dysregulation, gullibility, and other neurodevelopmental problems are intertwined with childhood experiences of abuse, neglect, trauma and loss. Like so many other residents of death row that colleagues in high stakes murder cases and I have evaluated, Jack's fight for the justice system to understand his intellectual and neurodevelopmental disabilities is still underway. His attorneys are appealing for reconsideration of the death penalty on grounds of intellectual disability (i.e., "mental retardation").

In these legal life or death battles, prosecutors and public defenders position their arguments in such a way to understand how certain facts provide mitigation for the crime. Akin to DNA evidence, the science to assist with ND-PAE diagnosis – quantitative electroencephalography (qEEG), magnetic resonance imaging of the brain (MRI), craniofacial morphometric measurements, intelligence testing including executive functions, and a host of other evaluative procedures help establish that prenatal alcohol caused the brain damage associated with ND-PAE.

We can do more to help the legal system understand these individuals. Certain provinces of Canada, as well as Washington State, Alaska, Minnesota, and a select few other progressive states have established "FASD Courts," for individuals with prenatal alcohol exposure and their families. In their meeting of the House of Delegates August 6-7, 2012, the American Bar Association adopted a resolution recommending training on ND-PAE (Fetal Alcohol Spectrum Disorder is the term they use):

> RESOLVED, That the American Bar Association urges attorneys and judges, state, local, and specialty bar associations, and law school clinical programs to help identify and respond effectively to Fetal Alcohol Spectrum Disorders (FASD) in children and adults, through training to enhance awareness of FASD and its impact on individuals in the child welfare, juvenile justice, and adult criminal justice systems and the value of collaboration with medical, mental health, and disability experts.
>
> "FURTHER RESOLVED, That the American Bar Association urges the passage of laws, and adoption of policies at all levels of government, that acknowledge and treat the effects of prenatal alcohol exposure and better assist individuals with FASD.
>
> ©2012 by the American Bar Association. Reprinted with permission. All rights reserved. This information or any or portion thereof may not be copied or disseminated in any form or by any means or stored in an electronic database or retrieval system without the express written consent of the American Bar Association.

Notwithstanding these intervention efforts, the focus of the national debate about the death penalty has been financial over the past few years, with one study revealing costs in Maryland of up to $186 million for five death row inmates who were awaiting execution in 2013. Based on my own experiences with these clients, in February 2013, I wrote letters urging Maryland state legislators to repeal the death penalty and was delighted when the state put an end to the arcane form of punishment. Criminology research, forensic data, and clinical experience indicate that there is a high probability that any one of the five men that were awaiting execution in Maryland may be suffering from undiagnosed brain damage due to ND-PAE. We must shift the paradigm of the debate toward a more enlightened, humanitarian direction based on neuroscience.

As of October 1, 2013, there were 3,088 death row inmates in the United States, with daily fluctuations from new convictions, appellate decisions, sentence commutations, deaths (through execution or otherwise), and exonerations. One of my close colleagues, Dr. Natalie

Novick Brown,[18] a clinical and forensic psychiatrist is following in the footsteps of her esteemed mentor, Dr. Ann Streissguth from the University of Washington at Seattle. Dr. Novick Brown has written countless academic journal articles and is an internationally recognized leader on the topic. In addition to seeing individuals involved in sexual offenses and other felonies, she has evaluated over 70 high stakes clients involved in *capital habeas* cases, either in a pretrial or appeal phase of their sentencing

While six percent of individuals commit about two-thirds of the violent crimes, a majority of those who perpetrate violent and persistent crime have both cognitive deficits and histories of witnessing or experiencing abuse during childhood. Since alcoholism is commonly found in homes with domestic violence and child abuse, many of the children have both ND-PAE (i.e., neurocognitive and other related problems) and have witnessed or experienced abuse. The unfortunate outcomes to society are both deadly and costly in individuals who have never received proper diagnosis or treatment for the condition.

In my travels to death row in states outside our own, the inmates I have evaluated have significant documented prenatal histories (by numerous affidavits in each case) of moderate to heavy, binge alcohol exposure. Each and every one of them was found to have some degree of brain damage resulting from maternal alcohol use. Their mothers often were from rural communities, had limited education, and drank with their husbands as a means of coping with poverty, domestic violence, and mental health issues of their own. One's ethnicity is not a protective factor, however, ND-PAE is found to be much higher in African American and Native American communities. Based on these experiences, it is my belief that all individuals on death row at least deserve to be evaluated to see whether they may have brain damage resulting from their mother's use of alcohol.

It would be tragic if anyone was put to death as a result of an act s/he committed that was primarily due to faulty neuronal wiring that predisposed that individual to a heightened stress response

---

[18] Brown NN. FASD Experts, Seattle, Washington. http://www.fasdexperts.com/.

(over-reacting in stressful situations), impulsive behavior, failure to understand consequences of one's actions (i.e., know right from wrong), or make sound/rational judgments. All of these problems can be seen in individuals with ND-PAE, who often have numerous difficulties with adaptive functioning, maintaining gainful employment, living independently without resorting to theft or prostitution, and long term significant relationships.

Like DNA evidence, a diagnosis of ND-PAE is grounded in science and sheds light on mitigating factors that may contribute to one's actions that lead to murder or any other unfortunate circumstance. Brain imaging studies, advanced electroencephalography (EEG), and other clinical methods are used in the medical evaluation of individuals suspected of ND-PAE. The individual's gullibility makes them easily led and influenced, to the point of being duped into a "murder for hire" without the cognitive wherewithal to understand the reality of the plan. A false bravado can lead them to perpetrate a crime without fully appreciating the consequences of their actions.

Meanwhile, we also have the financial burden at a time when our National government is struggling to stay open for business. In 2008, the Urban Institute examined 162 capital cases that were prosecuted between 1978 and 1999 and found that those cases cost $186 million more than if the death penalty not existed as a punishment. Each of the state's five executions cost Maryland taxpayers $37.2 million since the state reenacted the death penalty. A single death sentence costs an average of $3 million, including investigation, trial, appeals, and incarceration costs. That's about $1.9 million more than the cost of a single non-death penalty case, that is a case resulting in life in prison. At every phase of a case, according to the study, capital murder cases cost more than non-capital murder cases ...

The human rights tragedy of prenatal alcohol exposure should be a wake-up call for the disenfranchised and a call to action for the enlightened. With 2-5% of school age children having ND-PAE, more effort and financial resources should be focused on funding individualized, model programs to help children overcome their unique challenges and enjoy learning. Head Start Programs, day care centers, foster care, and other institutional settings should screen children for ND-PAE and provide the proper interventions – including parenting education, substance abuse treatment for parents, domestic violence programs, and other family supports.

If more widespread prevention and treatment initiatives were introduced in communities for all childbearing age populations, and women were afforded access to treatment and contraception prior to pregnancy, we would eliminate much of these pre-determinants to crime and violence. As it is, children with prenatal exposure to alcohol raised in stressful, physically violent homes then grow up, seeing the world as hostile and unsafe, having to fight their way through life. The interaction between witnessing or experiencing abuse and neurodevelopmental issues are associated with higher rates of violent and persistent antisocial behaviors, early diagnosis. However, appropriate interventions, and a supportive environment can be protective and a deterrent to criminal behavior.

Thus, the stage is set for individuals with undiagnosed ND-PAE being misunderstood - unable to be rehabilitated, lacking remorse, having poor judgment and lack of conscience, and needing to be locked up in a cell with the key thrown away. Perhaps I'm the eternal optimist, but I believe a paradigm shift in understanding will help stop the march of the penguins off the iceberg into death row. By identifying these children from an early age as neurodevelopmentally challenged but having as much potential and strengths as needed to live a happy, fulfilled, productive life, we can reframe their trajectory into one of hope.

Putting systems in place at an early age to provide an external brain and safety net for individuals with ND-PAE can prevent their proverbial march off the end of the iceberg into crime, homelessness, and mental illness, or possibly even death row. Identifying these children during infancy and toddlerhood will be the first step to putting appropriate systems in place to help treat and intervene with the affected children, their mothers and families. Perhaps then we can catch the Jack Fords and Adam Lanzas before they fall through the proverbial cracks in our fractured system of care.

The next chapter, *Jacob's Ladder – Climbing the Iceberg,* provides hope for patients in early, accurate diagnosis and treatment as well as unconditionally loving, tolerant, educated parents who fiercely advocate for their children to receive services and support.

Chapter 3

Jacob's Ladder – Climbing the Iceberg

12 And he saw in [his] sleep a ladder standing on the earth, and the top thereof touching heaven; and he saw God's angels going up and going down thereby ...

— Gen 28:12, Wycliffe's Old Testament.[19]

I first met Jacob in the Spring of 2010, just shy of his 8th birthday. Both tall and broad shouldered, he appeared the size of a well-developed 9 or 10-year-old boy – not much shorter than me – with striking Russian features resembling his adoptive dad in appearance and physique. His piercing hazel eyes darted around the room, soaking in the gestalt of my office. He asked, "You like Native Americans, don't you?" – noticing the Southwestern motif. My goal in that first session was to help Jacob feel comfortable and open to the experience, evaluating his body language, behaviors, movements, attitude, and "affect" (i.e., outward mood). Noting his perceptive nature, low tone posture, and open yet cautious attitude, I smiled and said – "Wow, Jacob! You really know how to figure things out about people."

We explored my games and toys, art supplies and doll house, Moon Sand, building blocks, farm animal figures, and Army men. I wondered out loud what types of things he likes to play with and he said he wanted

---

[19] Wycliffe's Old Testament, Translated by John Wycliffe and John Purvey. A modern-spelling edition of their 14th century Middle English translation, with an Introduction by Terence P. Noble (Editor and Publisher), 2001 and 2010.

to play with the sand and animals and Army men. This was to become his favorite activity during our sessions together – even as recently as his last appointment at age 13. As we set up the sand tray and figures, I told Jacob that I really love helping kids understand themselves and their parents better – like giving them an instruction manual. We talked about issues I've helped other children with – being afraid of the dark, not sleeping well, playing video games too much, having problems with school work, and sucking one's thumb as a pre-teen. Jacob enjoyed that part of the discussion, chiming in his own ideas about what types of issues I might help kids with. A fairly light-hearted exchange ensued complete with 4-year-old bathroom humor. Jacob's face lit up as he asked, "Do you help kids that pee on them self? Or can't sit still?"

Up until then, Jacob skirted directly around any references to himself, changing the subject or distracting himself further in the play activity. I tried countering his evasive tactics by talking about why he came to see me. All at once, he jumped up from the Moon Sand, pelleting me with a handful of sand and Army men. Aghast, I watched as he man-handled the oversized chair out from the wall and climbed behind it. He spent most of the session crouched under a makeshift tent draped over the chair. He vacillated between sucking his thumb and yelling, "I hate this place! I'm going to break this house! I want to go home!"

Being less savvy in those early days of my practice, I didn't appreciate that he had already given enough information and everything else was gravy. He had told me what his problems were in a way he could tolerate through play and indirect references. By his question about the patients I see, he told me he was unable to control his bladder and has problems staying in his seat at school. Asking him to admit he had those problems, I sabotaged the remainder of the session. By confronting his evasion tactics, I unwittingly flipped a switch that triggered a series of melting circuits in his brain.

Reviewing the notes from his parents' session the day before, the tantrum that ensued was a glimpse into behaviors that led to the appointment with me. It was clear that a gentle approach was needed to reach the frightened boy on the other side of the arm chair. I imagined that for him, I was just another stranger asking him to talk about very difficult, deep emotions – painful even for many typically developing children. My pressure to talk about his undesirable behaviors pushed him to admit his nightmarish Groundhog Day life. Reflecting on the outcome of our discussion, I interpreted his response to the stress as anxiety – a "fight or flight" reaction. Feelings of insecurity and discomfort triggered anxiety, which was at the root of his regressed, toddler-like behaviors.

Though I was just getting to know Jacob, his parents, Dave and Karen West, had experienced his low frustration tolerance and easily triggered flash point from an early age. We had met the day prior to discuss their concerns and perspectives about their son. In that first visit, I learned that Jacob was adopted at 15 months from Russia and was the size of a gaunt, frail six months old baby. Jacob had been in a Russian "baby home" since birth – having been placed there because his alcoholic mother was found unfit by the government child welfare agency. The West' knew that Jacob had a number of medical issues – prematurity, low birth weight, failure to thrive, and developmental delay which was said to be due to lack of stimulation. He was left in the crib at the Russian orphanage for such extended periods of time that he developed torticollis, a chronic muscle spasm of the neck causing his head to tilt to one side. After arriving in the US, he began to receive occupational therapy services for torticollis, fine and gross motor problems, and sensori-integration issues, with limited to minimal improvement. He had attended a private day care and preschool program, had to repeat Kindergarten at the local elementary school, and was failing first grade by the time he saw me. The West' believed that their son struggled on a number of different levels but had been told by his school that he was capable of "on grade level work."

A picture of his lonely, distant eyes sitting on his mother's lap during the long series of plane rides to the United States was hauntingly familiar. During my fellowship in child psychiatry, I had seen films about very young children in post-World War II orphanages with that glazed appearance who had institutional depression from profound neglect and despair. At Children's National Medical Center, I had learned and read a number of papers about problems with attachment to parents caused by such early experiences. When a baby's cries are not responded to, over time, they give up trying to get their needs met. Scientists have created a number of experimental animal models of "learned helplessness" – an environmentally-induced form of depression. Pavlov tied dogs to shock pads and electrocuted them despite their cries of pain. When he finally unchained them, most remained on the shock pads, enduring the intermittent shocking. Other researchers have replicated the model in various species. If rats are repeatedly placed in a deep aquarium of water with no way to escape, many will try in vane until they are so exhausted that they give up trying.

Transitions had always been hard for Jacob, especially around loud noises and crowded settings. These seemingly minor triggers tend to unravel Jacob, leading to him "fighting or fleeing" to get out of the situation. He also had difficulty appreciating the boundaries of personal space, leading sometimes to others getting hurt. For example, swinging his book bag around at the bus stop would often lead to him accidentally hitting another child, who felt he did it on purpose. He would be scolded by the teacher or bus line attendant, or even sometimes sent home with a note. Such insignificant events in Jacob's life led to catastrophic outcomes, such as his parents getting called by the principal and having a school meeting. Jacob would feel so anxious and shamed that he would have a meltdown when he got home that day.

Since Jacob is a very sensitive and perceptive boy who is eager to please his parents and other adults, their reprimands lead to him feeling he is a "bad boy" and different than other kids his age. On some occasions, his parents would get a glimpse into his internal conflict between an unconscious yearning to behave and the "beast from within" struggling to be heard. He would say, "I like Miss ____, she helps me learn ... I like reading, science, math, art, music, or gym ... I don't like getting in trouble, but when I get mad, I can't control myself ... I know it's my fault when I get in trouble ... I want to have a good day ... I want to be with my sister and go to 'Bar-T,' [her aftercare program] ... I want to read to Mr. T or Miss F [teacher aides]."

Some months earlier, Jacob had been evaluated at specialty clinic at one of the local university teaching hospitals. He was diagnosed with Fetal Alcohol Syndrome (FAS), based on his facial features, a history of prematurity and low birth weight, and developmental delays that persisted even after several years of therapy and educational supports. The reports recommended additional neuropsychological testing and interventions, eliminating one of the two medications he was taking, and changing some of the West's parenting strategies. Because she was already feeling overwhelmed and incompetent to parent Jacob, Karen felt blamed for what she and Dave "didn't do" right in their parenting style. The changes they made to the medication led to worsening behaviors and him biting someone at school, so they decided to seek advice from a private psychiatrist closer to home. That psychiatrist had increased Aripiprazole, abruptly switched the stimulant he had been taking to a completely different medication, and added a third drug for behavioral regulation. His parents, Dave and Karen, said he was transformed from a "model first grade student" with few discipline problems, into a disruptive, confrontational, oppositional child – resulting in his being sent home from school repeatedly for dangerous, unsafe behaviors. Karen felt that the increase in his defiance and belligerence coincided with the medication change about 4 weeks prior.

At school, Jacob had become physically aggressive toward himself, teachers, and classmates – quickly becoming defiant and disrespectful. In the afternoons, he was refusing to do work, telling his teachers what he would or wouldn't do (such as getting on the bus for after care, exiting the room to attend a special, completing his work, listening to the daily routine in the morning, following the teacher's directions). Often, this oppositionality would lead to loud, verbal confrontations with the teacher in which Jacob would rip up his work, draw on his desk, and/or shout that he is "not smart," or "I'm stupid!" or "I'm a baby!" or "I'm a dummy!" He would begin scratching his arms, face and forehead and to focus on who and what he "hates." He made frequent, threatening statements – "I will push [item] off your desk!" "You need to shut your mouth and quit talking!" "I am going to scratch my face!" "I am going to the lunchroom and start a food fight." At a loss for dealing with what they viewed as "oppositional behaviors," the teachers were attempting to provide one-on-one instruction, time outs, independent areas to work, frequent visits to the office, phone calls home, and withholding rewards.

At home, he was demonstrating physically regressive behaviors – curling up in a ball on the floor in a fetal position, kicking to the point of his shoes coming off, and having lack of eye contact or attention to others that were speaking to him. Karen and Dave described Jacob's frequent mood shifts, switching from "attentive to disruptive in a flash," talking about a variety of random topics, angry, argumentative, and rageful. In this negative frame of mind, he would shout about a variety of people – "I hate my teachers … students … the principal … school!" "I hate my mom … sister … myself … all Jewish people!" He would also say things like "I want to turn into a giant and crush the school and burn it down!" The West's and their adoptive daughter were traumatized by his rages, which often erupted into holes being beat into the walls and kicking in the front door of the house at the young age of 7. These periods of upset would take anywhere from 30 to 40 minutes to calm down from and get back on task at school. He often would complain afterwards, "my head hurts" and "I can't control how I act" (i.e., that he couldn't control himself during the episode), and that he didn't want to get in trouble. Often, he would yell, "I want my reward!!!! Now!!" His behaviors transformed into whining, making crying noises, and even farting in the aftermath of one of these explosive episodes. Despite the intensity of the incident, he rarely would apologize for his behaviors and would move on as though nothing had happened. Dave, Karen, his sister Miya, and anyone around him would feel terrified during the events and traumatized afterwards. Miya developed a severe anxiety reaction called acute stress disorder and often hid out in her bedroom with the door locked until he settled down. Jacob often confabulated to staff at school that his parents would lock him out of the house or physically harm him by kicking or slapping – often acting out the action to whomever was listening.

His classroom disruptions affected his learning, changing the teacher's dynamic and peer interactions based on their misperceptions of his intent. Based on school reports and discussions with teachers and administrators, Jacob was performing on grade level and therefore did not qualify for services at school. In fact, as I later pointed out in an educational management team meeting at his school, though he failed kindergarten twice and was nearly finished with first grade, his performance was barely at the level of the first semester of kindergarten. He was perceptive enough to recognize that other children finished assignments much faster than he did, knew the answers when they raised their hand, and were reading and completing math problems accurately, and were writing legibly and neatly. In turn, Jacob felt uncomfortable, frustrated and misunderstood at school and unable to please his parents at home. Daily negative reports from school and his spiraling out of control at home left Dave and Karen feeling inept and hopeless to help their child. At a crossroads in understanding his challenges, they began to lower their own expectations for him emotionally and academically, recognizing that they needed different strategies to parent. They understood that he needed more supports to cope on a daily basis for his behaviors and academics. They were concerned that the school was missing the "bigger picture" of his lack of academic progress and accomplishments. They also worried that his demeanor and appearance (scratches all over his face from his nails, pencil leads, crayon) would frighten other children and effect his relationships with them.

Many psychiatrists who see children like Jacob are overworked and have long waiting lists, leading to limited time for management of patients between visits. Early on in my practice, I decided I would only take one or two new patients for evaluation and treatment per month in order to provide intensive care for them as they transitioned off cocktails of medications. Though I have become more seasoned and able to manage greater numbers of patients, I still believe that a high level of hands on care is necessary to help these patients and their families. Such intensive care is not realistic in an insurance model that does not reimburse for telephone calls or school visits. This was the system the West's were in with both their children. Their plight eventually led to a consultation with me.

Karen had learned about my practice through the National Organization on Fetal Alcohol Syndrome (NOFAS)[20] – a nonprofit

---

[20] www.NOFAS.org

advocacy group for prevention and education about Fetal Alcohol Spectrum Disorder (FASD). She had cancelled the first appointment with me after finding Dr. Ronald Federici through an adoptive parent support network. His book published in 1998, *Hope for the Helpless Child: A Guide for Families,* had caught her attention – giving her a glimmer of hope that there was something she could do as Jacob's parent. Dr. Federici, who has since become a close colleague, practices on a farm just outside Clifton, VA. He has adopted eight children from Russia and Romania and raised several U.S. foster children. World renowned for his work with post-institutionalized children, his efforts have led to a program in Romania called the Bucharest Early Intervention Project. The program has shown promise in a variety of areas for children placed in supportive, loving foster homes with nurturing, caring parents prior to the age of 24 months. The parents receive specialized training and assistance from community social workers with expertise in working with children from orphanages who have special needs, trauma histories, and other issues. The program has found that appropriately trained caregivers and many other early interventions before the age of 2 can improve the child's ability to bond with adoptive parents and have fewer behavioral issues by school age.

Karen eventually called me back to reschedule after learning that Dr. Federici is a developmental neuropsychologist, not a child psychiatrist. While he could provide invaluable neuropsychological testing, he would not be able to provide both medication management and long term therapy. Instead of interpreting her call back as "doctor shopping" or as a sign of ambivalence, experience told me otherwise. She was climbing up and down the ladder of specialists in our fractured system of care seeking the proverbial "guardian angel" for her child. Jacob's ladder was like so many desperate others I saw frequently in my practice. His laundry list of diagnoses and out of control behaviors would leave most parents feeling overwhelmed, as though in a revolving door of Groundhog Day. Once they rescheduled with me, Karen emailed me during a crisis at his public elementary school, stating "they are worried for his safety as well as others in the school ... [he] has been experiencing extreme frustration, rage, anger, loss of interest in most activities, threatening behavior, scratching himself." Compassion led me to see them later that same day instead of the following week.

From his appearance, Jacob did not fit the textbook criteria of a frail child with "dysmorphic" (abnormal) facial features of prenatal alcohol exposure. Though he was barely fitting into 6-month clothing at 18 months, he had eaten well and gained weight rapidly after being adopted. He also gained 10-15 pounds on the antipsychotic Aripiprazole under the care of a different psychiatrist. By the time of my initial evaluation, Jacob's lab work showed he had developed high cholesterol and was in the early stages of "metabolic syndrome" – a pre-diabetes condition caused by obesity. The relatively moderate doses of antipsychotic medications had caused weight gain, leading to the other medical complications. During other blood work down the road, I learned that Jacob's liver enzymes were defective in digesting certain drugs – likely caused by changes during development from chronic exposure to alcohol. In medical terms, "epigenetic changes" can occur in the developing fetus in such a way to cause drugs and chemicals to be processed differently by the liver. Because of his slow metabolism of the antipsychotic medicines, the drugs cause worse side effects than in people with normal liver enzymes.

Prenatal alcohol exposure also caused Jacob's brain to be wired differently than a typically developing child. Like a jerry-rigged 1812 manor house with 1940s electricity installed in the late 1930s, Jacob's brain had connections that were affected by switches not biologically programmed to control those areas. During their migration in early pregnancy, brain cells lost their ability to plot their course and steer in the right direction, leaving them in places they were never intended to develop and/or dead from the effects of alcohol. Other areas have frayed wires like mice had chewed through the wiring.

The "amygdala" can be considered the emotional thermostat that regulates the brain's temperature and turns on or off the brain's "furnace" when necessary. I often point out the "panic button" deep in the brain using a model. In addition to functioning as an "on – off" switch and thermos-regulatory function, the amygdala is also responsible for the intensity of the reaction, much like a rheostat adjusting the voltage and intensity of the reaction to a stimulus. Jacob's amygdala is faulty – leading his furnace to turn on at inappropriate times and not sensing when he was too angry (volatile) for the situation. A fairly benign trigger leads his faulty wiring to see the situation as dangerous and take extreme measures to avoid it. Sometimes his frayed wires send sparks off to trigger other neurons to fire. His circuits overheat and he goes into "meltdown mode." Another analogy is having "crossed wiring" in an old house – knowing which switch may inadvertently turn on a circuit in another part of the house, leading to smoldering wires. Parents of these children learn to sense all too well the moods and behaviors that "smell like" an electrical fire is coming. More details about this can be found in Appendix A-1.

Jacob's wiring is also a little like my 1965 brick ranch house built with plaster walls, not dry wall. Certain sections of the house have been rewired over the years as the former owners put electrical outlets in different places than the original builders. When I have both my coffee pot running and my microwave – which are plugged into the same outlet, the circuit overloads, shutting off the breaker. So nearly every morning for the first several weeks in the house, breakfast would be interrupted to turn back on the breaker. Parents learn over time the triggers for circuit overload, understanding that one activity at a time helps avoid power failures and meltdowns.

Jacob's poorly wired nervous system crippled his ability to function in a world tolerable to most 6 to 8-year-old children. His low frustration tolerance, lack of social skills, limited cognitive abilities, and hypersensitivities to noise, smells and textures left holes in his basic life skills. His emotional, cognitive, language and social skills were more consistent with a 2 to 3-year-old child than the 8-year-old boy in front of me. The smoldering, frayed wires left his parents and sister walking on egg shells most of the time, knowing than an electrical fire was brewing.

Like other individuals with ND-PAE, I hypothesized that Jacob had brain damage from prenatal alcohol exposure. As international adoption studies had shown, approximately 50% of children adopted from the former Soviet Union had evidence of some degree of neurodevelopmental issues associated with prenatal alcohol exposure. In my practice, I had found evidence of "static encephalopathy," "epileptiform encephalopathy" and "epileptic encephalopathy" in children from international adoptions. These three types of brain damage are associated with varying degrees of abnormal brain wave activity on electroencephalography, or EEG. In neurology and psychiatry, we view these conditions as associated with brain irritability. Oftentimes, we can see evidence of the brain damage on magnetic resonance (MRI). I have detailed more information about diagnosis and treatment of these conditions in Part 2 of the book. Believing that Jacob's mood outbursts were due to abnormal circuits causing a cascade of electrical impulses traveling to a variety of brain areas causing over-reaction to fairly benign stimuli. I sent him for both an EEG and a MRI of his brain in order to better understand what was happening. I also ordered a variety of blood tests, including genetic testing of his liver enzymes to determine how he metabolized psychiatric medications and his serotonin receptor and transporter genotypes to predict his response to selective serotonin reuptake inhibitors.

As I predicted, his sleep deprived and 24-hour EEG resulted in "irritability" or abnormal brain wave activity, while his MRI findings showed an abnormal corpus collosum – the bundle of neurons connecting his two hemispheres. These studies, combined with the neuropsychologist's report, helped the school understand that his behaviors were a result of his cognitive and neurodevelopmental limitations. The abnormal EEG helped me understand that a mild anti-seizure medication might be the best option for improving Jacob's symptoms. Over a series of many weeks, I switched him to lamotrigine and tapered off a number of the other medications. I also added clonidine to help his impulsivity and hyperactivity and to minimize the need for stimulant medication. The changes required a few different hospitalizations, namely to be able to control his rages while the medications were being tapered.

Ultimately, after several failed attempts to place Jacob in a variety of public school settings, traumatizing him each time by causing more emotional outbursts and ineffective, punitive consequences, the county board of education professionals eventually placed him in a therapeutic school setting for children with severe learning disabilities. There he spent the next four years, progressing only as far as the middle of third grade in his academic goals. By the time he was nearly 13, he was still at the fifth grade on paper but had not progressed on his Individualized Education Plan goals in the past year. At that point, we reviewed the need for placement once again, understanding that a therapeutic residential school would be the most appropriate given his IQ of 56 and lack of even the most basic adaptive functioning skills.

Adaptive functioning is a measure of one's abilities to navigate social situations, complete daily activities such as hygiene and meal preparation, and perform academic tasks. Individuals with ND-PAE tend to have lower adaptive functioning compared with their intellectual quotient (IQ). In Jacob's case, unlike other 8-year-old children, he was unable to shower himself in an age appropriate manner without significant assistance from his parents, to take care of his hygiene on toileting, and to prepare cereal for his breakfast. From a social skills perspective, he was vastly immature compared with his typically developing peers – not understanding nonverbal cues, being gullible and easily influenced, and overly trusting of strangers. Academically, he was far behind his peers in math, writing, and reading despite several years in a therapeutic school. I believed that his IQ of 56 was an accurate assessment of his intellect and that his even more profound adaptive deficits prevented him from being able to navigate through his environment in school and at home without significant upset.

Until school age, his parents were unaware of the degree of his disability, having been told when they adopted him that he was growth deficient and had health problems because he had been in an orphanage for the first 15 months of life. The adoption agency said he would "catch up" on his milestones and gain weight once he had gotten proper love and nutrition. The West's believed that their son was a survivor – possibly more resilient than most other children in the same Russian orphanage and they were determined to help him thrive by giving him a loving, nurturing home environment as they had been told would help him overcome his harsh beginnings.

The details of the West's story – that of going to Russia to adopt one child, who they had read about and seen a photo of on the adoption agency's dossier, and coming back with a different boy as well as a girl, was becoming all too familiar. The book *When Rain Hurts* by Mary Greene epitomizes that story, though in their case they went to Russia with the intention of adopting a girl, who was not available when they got there, so they were given a different child and a boy, who they felt pressured to take if they wanted the girl. I have since learned that this common phenomenon among international adoptions from the former Eastern Bloc was among many reasons that The Hague Convention recently established guidelines for adoption agencies to follow with oversight and regulation. In 2010, Russia temporarily halted adoptions to the United States due to backlash after Torry Hansen, a single mother from Tennessee, sent her adopted 7-year-old son back to Russia alone with a note saying that she could not parent him due to multiple psychological issues. Like the West's, Mary Greene and her husband, and Hansen, American families felt deceived by adoption agencies that provided limited or misleading information on the children they were adopting. Many of the families paid a large amount of money for the children, including additional services to have videos and photographs of the children reviewed by physicians who specialize in international adoptions.

The parents of these children were ill-prepared to handle the multiple neurodevelopmental and neuropsychiatric challenges the children faced. Most describe the preparation from the adoption agencies as standard parenting classes designed to orient typically developing youngsters into the U.S. culture and a family lifestyle. Whether agencies themselves have been aware of the children's problems remains to be seen; however, translations of medical documents and paperwork indicate complex neurodevelopmental issues such as perinatal encephalopathy, syndrome of motor dysfunctions, vegetovisceral dysfunctions, encephalitis, organic encephalopathy, severe developmental delay to other terms that are either arcane or obsolete in the medical dictionary. In cases where information was made available to the parents, adoption professionals told them the children were perfectly healthy but the only way to adopt them was to record them as "feeble" or "weak."

The West's remember bringing Jacob home with legs so weak he could not stand at 15 months, being too small for size 6-month baby clothes and exhausting them at mealtimes with his anxious feeding habits – gulping down mouthfuls of food as though he would not have another meal. The past 4 ½ years we have worked together have resulted in better understanding of what was not as obvious all those years ago - that Jacob requires 24/7 supervision, will likely never live independently, will require significant vocational support throughout the duration of his employment in even unskilled, menial jobs. In an educational system designed for children who are college bound, there are few tracks appropriate for children with multiple developmental disabilities such as Jacob, even if they are able-bodied and appear capable of honest, hard work.

It has been nearly five years since I met Jacob. I have seen him through a total of 6 hospitalizations, four of which were within the first few months of seeing him to "wash out" the medications and treat the metabolic syndrome turning his body to mush. Ironically, the best and worst outcomes were his only admissions to a small community psychiatric hospital with a Russian-trained physician leading his treatment team. The first time under her care, she seemed to understand that his medications were exacerbating his emotional instability. She selected a relatively benign atypical antipsychotic to help with self-regulation. However, since that time, I had completed a panel of genetic tests to determine his metabolic status for a broad range of psychiatric meds (CYP-450 genotyping – described in Part 2) and his serotonin transporter and receptor genotype in order to predict his response to pharmaceutical drugs. As a result, I found that he is a very slow to non-metabolizer of certain other antipsychotic medications and tried to tell the physician that on his second admission four years later. That was to become the time Jacob almost died from a doctor's medical decision.

On admission to the hospital, we knew that Jacob's medications needed adjusting. He had grown substantially since his last adjustment – taller and leaner, now in the throes of puberty. His adolescent brain was having neurons come on line in places they were never intended since they had lost their signaling along the way. We knew he had significant visible evidence of prenatal alcohol exposure on MRI since his corpus collosum was malformed, as were the grooves and pits of his cerebral cortex. With the natural, developmentally appropriate hormonal changes and fluctuations in neurotransmitters in his brain, Jacob had started to once again become oppositional, not following directions, hitting or running away from staff at school, and not engaging appropriately at home with his sister and parents.

During the hospitalization, we agreed to taper him off one of his medications, an antipsychotic, and begin a different medication. We discussed his neurotransmitter systems and how his CYP 450 enzyme system was defective (not metabolizing certain medications). The antipsychotic he had been started on during his initial hospitalization under her care was not metabolized by that system and therefore had not caused unexpected side effects. We hoped his symptoms would be better treated with a different medication at a minimal dose. The psychiatrist agreed to gradually taper the medication and begin a low dose of another antipsychotic. Because she felt pressure from the insurance company, she quickly withdrew the first medication and started a second antipsychotic. When he began to show signs of restlessness which she interpreted as worsening hyperactivity and impulsivity of ADHD, she increased his stimulant and the new antipsychotic. Unfortunately, both those are metabolized by the enzyme that is minimally to non-functioning in Jacob's liver.

That weekend, his parents called me, distraught and concerned about his physical condition. They described him as out of control - up and down, pacing back and forth, flighty, manic, and giddy. Listening to their description of his behaviors and eliciting further symptoms of rocking, fidgetiness, hair twirling, and inability to sit or stand still, I formulated an impression. The resulting increase in the drugs led to a condition called akasthesia – a severe side effect of high dose antipsychotic medications. The reason I believed Jacob had developed these symptoms from a modest dose of antipsychotic was due to his faulty enzyme system and competition between the antipsychotic and stimulant for the limited amount of active enzyme in his system.

Incorrectly believing the second antipsychotic was not working, the psychiatrist decided to "cross taper" to a different antipsychotic, which meant starting a third antipsychotic while lowering the dose of the second. In this way, his liver enzyme system would have been further stressed, potentially resulting in a condition called neuroleptic malignant syndrome – a severe, catastrophic effect of antipsychotic medications. We knew from his developing rapid weight gain and borderline diabetes (metabolic syndrome) at age 8 that he had a history of extreme side effects from antipsychotics. NMS results in high fever, stiffness of the muscles, altered mental status, and problems with the involuntary (autonomic) nervous system - causing wide swings in blood pressure, excessive sweating and secretion of saliva.

It turns out that a neighbor's son had died from the condition at a residential school several years earlier. The young man suffered from a severe form of autistic disorder and had developed flu-like symptoms at the school. The nursing staff had sent him home, not recognizing he had progressively worsening symptoms from high dose antipsychotic medication. Though there was a psychiatrist on contract to the school two days a week, he had concurred with the nursing staff's assessment that the boy was suffering from an infectious illness. As a result of severe NMS, he died during the night shortly after returning home.

That insight and a passion for problem solving led me to construct a summary of my interpretation and clinical perspectives of Jacob's symptoms, which I faxed to the attending physician at the hospital and to the insurance company. I then reached out to a neurodevelopmental unit specializing in children and adolescents with issues such as ND-PAE. Through coordination of care, we transferred Jacob to the unit to safely washout the medications and stabilize his neurobehavioral issues. The child psychiatrist there was adequately trained in neurodevelopmental issues and accepted the philosophy that "more medication is not necessarily better." Following titration from all medications except the seizure control medication, we were able to transition him to a residential school - ironically, the same school my neighbor's son attended just prior to his death. Knowing that history, I had insisted years earlier that Jacob not go to that school due to limited psychiatric oversight. However, the reality in psychiatry is limited availability of residential facilities for severe neurodevelopmental issues. After Jacob's mother and the director of his previous school had toured the residential school and shared their excitement about the programs there, I agreed that it would be a suitable place for him. The school assured us that I could continue having regular office visits with Jacob to manage his medications.

During a medication checkup and parent guidance session just prior to his admission to the residential school, we spoke about my crystallizing plan to develop a therapeutic school and community in Maryland for kids with ND-PAE. Jacob quickly began telling me many things he would like to include on the property – a go cart track, a Native American camp site and tool making center, a tree house, and on and on. I gave him several sheets of white paper and a pen and he went outside with instructions to sketch his ideas on paper. Over the next 45 minutes, the boy who had been so out of control in my office those years earlier sat calmly drawing and labeling a "blue print" for the community farm and school he and I had talked about so many times before. "When can I move there?" he said as he proudly handed me his finished product. "I'm giving myself two years, Jacob. I'm hoping you can learn the things about what works and what doesn't for you at your new school so you can help make my school better."

Within the first few weeks at the school, Jacob developed a faint rash on his arms and legs. His parents were not informed by the school and only found out on his return for Thanksgiving break. We discussed the possibility of dry skin, hives, and other dermatologic issues. When he returned home for winter break, the rash had worsened to severe, diffuse hives and bull's-eye eruptions, many oozing and infected from his scratching. It was so severe under his armpits that he had limited hair left and the top layer of skin was completely sloughed off. The staff at the hospital did not believe it was any of their products or the anti-seizure medication; however, his mother suspected it was an allergic reaction to mold or a chemical in his environment. His mother sent back dye and perfume free products for them to use and instructions for someone to observe him showering and rinsing the shampoo and soap from his body. I concentrated on the possibility that it was a delayed reaction to the anti-seizure medication since it can cause a rash with that pattern of distribution. We immediately began two antihistamines and tapered him off the seizure medication. His rash gradually resolved and he continues to be doing well despite being on no medication.

Like other families of children with prenatal alcohol exposure, the West's have sought complementary and alternative methods as adjuncts to traditional allopathic medicine to help their children's fragile nervous systems. These methods have included neurofeedback and nutrition. The West's sought out a specialist in quantitative electroencephalography (qEEG) who uses a computer program hooked to brain electrodes to help rewire the brain. Over several years, they also worked with a primary care specialist in treating children with complex neurodevelopmental histories who found their vitamin and trace mineral profile to be similar to that of a chronic alcoholic. Researchers know very little about the gastrointestinal tracks of these children; however, a "leaky gut" phenomenon may lead to poor absorption of important minerals and vitamins. By treating the underlying nutritional deficiencies, the treating physician believes he is able to improve functional capacities (emotional regulation, insights into how their behaviors affect others, and "triggers" for their anger outbursts). We know that these children suffer from tremendous growth problems due to hormonal dysregulation (thyroid, reproductive, and growth hormones). These hormone imbalances are thought to account for some of the growth deficiencies, increased weight gain in pubescent females, and other metabolic and reproductive changes. In Jacob's case, he had gained a fair amount of weight while on an atypical antipsychotic before beginning to see me.

Today, reviewing his history and looking back on how far he's come, his mother and I agree that the history depicted above omits the intensity of his violent rages, uncontrollable crying spells, and other significant emotional problems during his early years. I firmly believe that Jacob's current living situation in a pastoral environment learning life skills rather than being in a diploma track is better suited to his neurodevelopmental level. He is proud of working in the cafeteria where he is learning to assist the staff with preparing simple meals, setting the tables, prepping vegetables, and doing minimal chores in lieu of the pressure toward high level math or reading novels.

Jacob, like the other 60,000 or so Eastern European adoptees, was given hope in the form of Karen and Dave West. Blindly, without knowing any of his history or the impact of the first 15 months of his life, they were as angels lifting him out of isolation of a Russian baby home, providing him a comparable as the Pharaoh's daughter lifted Moses from the reed basket centuries ago. Their courage and determination struggling up and down Jacob's biblical ladder has been his saving grace. In his dream of the ladder with angels ascending and descending into Heaven, God said to Jacob - "I am with you and will watch over you wherever you go, and I will bring you back to this land. I will not leave you until I have done what I have promised you." Echoing this sentiment to Jacob, given our relative capacity for living in society and his need for protection and care, I feel we all owe him and so many others affected by our social drug of choice to nurture and support rather than penalize and persecute by criminalizing their maladaptive behaviors – an artifact of not adequately protecting them from their own social, emotional, and intellectual vulnerabilities.

Understanding the extent of this silent epidemic will assist in preventing future generations of brain damaged children.

Chapter 4

THE SILENT EPIDEMIC

Historical references elucidate the argument that the government, medical community, and alcohol industry are remiss in their duty to warn the American consumer. — SDR

In today's society, an epidemic of hidden birth defects is happening silently to the minds of our unborn. How can we be so blind and numb to prenatal alcohol exposure? By turning a deaf ear, we have metaphorically thrown out the baby with the "fire water." Unlike epidemics of contagious diseases, ND-PAE begins well before a woman may know she is pregnant. Our love affair with alcohol has hidden this dirty little secret in the rural, urban, and suburban areas of our great nation. Overcoming this veil of silence and its impact on unknowing and innocent individuals will be similar to enlightening the ignorance that perpetuated Bubonic Plague in the fourteenth century and illnesses such as HIV and measles in modern times. An understanding of terms used to characterize illnesses of similar magnitude bolsters the argument that ND-PAE is an epidemic.

Historically, the term *epidemic* describes highly prevalent *contagious* (communicable) diseases brought into a community or spreading rapidly. More loosely, any disorder identified in a population at highly prevalent rates is considered *"epidemic"* in proportion. Today, *epidemic* is used to describe other conditions brought on by modern lifestyle behaviors, such as obesity, heart disease, and emphysema. The excerpt from *What Constitutes an Epidemic?* by Dr. Benjamin

Lee featured in Figure 5 was published in the late 19th century when epidemics of cholera, measles, tuberculosis, and other *communicable* (contagious) diseases still ravaged our communities.

Figure 3. What Constitutes an Epidemic?

...to determine just what proportion of any given population should be attacked in order to justify the declaration that an epidemic exists ... The answer given was that the disease should be declared epidemic when the number of cases should reach these proportions:

- *For a population of 100: 5%.*
- *For a population of 500: 4%.*
- *For a population of 2,000 to 5,000: 22 per 1,000.*
- *For a population of 6,000 to 10,000: 16 per 1,000.*
- *For a population of 20,000 to 50,000: 8 per 10,000.*
- *For a population of 50,000 to 100,000: 4 per 10,000.*
- *For a population of 200,000 – 1 per 10,000.*[21]

Once *sentinel cases* (initial affected individuals) of an epidemic are identified, scientists then uncover the *pathogenesis* (the way a disease develops) to better understand the *etiology* (underlying causes) of the illness. Public health professionals, researchers and physicians work in a coordinated manner to develop effective *treatments* (e.g., medications) through clinical trials; *interventions* (e.g., quarantine of affected individuals) to reduce the spread of illness, and *prevention* (e.g., vaccination) programs to limit further population exposure. Eventually, some conditions may become dormant in a given population – existing at a low, undetectable level *(endemic rate)* until the right environmental factors lead to an outbreak *(resurgence)* of the illness.

Before understanding that germs caused contagious illnesses, European and Mediterranean populations were decimated by transmission of an invisible, unseen yet deadly bacteria carried by rodents. Thus, Black Death spread as though by a curse, hidden, unseen, and unchecked from rats travelling by ship into Sicily and

---

[21] Excerpted from "What Constitutes an Epidemic?" by Benjamin Lee, MD; *Public Health Paper Report.* 1898; 24:97-99.

Constantinople on the silk route from China in the Middle Ages. Between 1347 and 1353, one-third of the European population was wiped out by the painful, hideous illness, ushering in the Dark Age of Europe. In 1894, during an epidemic of plague in Hong Kong, Dr. Alexandre Yersin, a Swiss/French physician and bacteriologist from the Pasteur Institute, linked *Y. pestis* with plague. He used Pasteur's methods to isolate the culprit: a Gram-negative, rod-shaped microscopic bacteria, infectious to both humans and animals. Thus, the germ theory led to discoveries of microbes causing diseases. Despite stringent public health methods, a resurgence of *Y. pestis* (the bacteria causing Black Plague) has happened in certain areas of the world in modern times. In turn, public health methods, such as sanitation, quarantine, and vaccination contained outbreaks of Bubonic Plague, and eventually cholera, measles, small pox, rubella, and tuberculosis. Each year, the World Health Organization reports thousands of Plague victims, who have much better prognosis with advances in medications for treatment.

Sentinel cases are the first cases identified for a given illness that signal the onset of an epidemic. For the AIDS epidemic, the *sentinel* cases were homosexual men developing life-threatening weight loss and infections. Public health professionals, scientists, and medical doctors working together characterized the population affected: young gay men frequenting communal bath houses in large cities such as New York and San Francisco, all with very low T-cell (white blood cell) counts. Other aspects of the pathogenesis became clearer when the cause of the disease was found to be a virus (Human Immunodeficiency Virus) brought to the United States from Africa. Eventually, it was found that the virus could be transmitted by sexual contact, intravenous needles used by drug abusers, blood transfusions in hospitals, and other means in which body fluids are exchanged between people. Drs. Elion and Hitchens from Burroughs Wellcome Company developed the first treatment for HIV/AIDS – AZT (or, azidovudine), an inhibitor of the reverse transcriptase enzyme that helps the virus multiply in cells. The unprecedented success of the drug and the need for containment of the virus led to the Food and Drug Administration "unblinding" the clinical trial and allowing all study participants receive the active medication instead of placebo.

Similarly, measles, existing at endemic rates in the population, exemplifies the phenomenon of *resurgence*. A few sentinel cases in a preschool, boarding school, or college signals a resurgence of the condition where children/adolescents/young adults are in close contact. Cautious public health measures, such as quarantining infected individuals and their contacts and ensuring vaccination prior to college admission, help limit spread and prevent future outbreaks.

Like contagious illnesses, ND-PAE has resurgence when poverty increases or social systems fail – as in the gin epidemic in Europe in the mid-1700s, the post-Civil War era leading up to the Industrial Revolution, and the collapse of the former Soviet Union. Between episodes of large scale "outbreaks," the condition remains relatively silent (endemic) in the population – misunderstood and hidden amidst the mental health and social problems of minority groups, socially disenfranchised communities, and the marginalized under class. It is an "out of sight, out of mind" phenomenon.

For ND-PAE, *sentinel* cases were initially most easily recognized from births to alcoholic women and described in the papers written by Dr. Lemoine in France in 1968 and Drs. Jones and Smith in 1973. Many of our prevention and intervention programs are based on these early findings (i.e., providing greater access to prenatal care for pregnant women, screening, referral and treatment of pregnant women for alcohol abuse, and a warning label advising pregnant women not to drink alcohol. Yet we have known since Dr. Kathleen Sulik's seminal paper in 1981 that ND-PAE occurs as early as three weeks after conception – well before most women know they are pregnant.

So, are we doing enough to educate the public about the reproductive health outcomes of alcohol?

In order to answer this question, let's examine society's response to a more obvious tragedy caused by the anti-nauseant thalidomide. During the late 1950s and early 1960s, about 12,000 Canadian and European infants were born with a constellation of malformations, including missing limbs and other birth defects.[22] The culprit was the anti-nauseant thalidomide—a teratogen that disrupts a number of processes, such as long bone development.[23] These defects were so horrific that developed countries banned thalidomide until recent months. Thalidomide caused an epidemic of babies in Europe and other countries around the world to be born with small "flippers" for limbs. This disaster happened because the pharmaceutical drug, thalidomide, is a chemical that causes birth defects (teratogen). Alcohol is also a teratogen, chemically affecting brain and nerve development (i.e., "neurodevelopmental teratogen") more than physical structures. Unlike alcohol, therapeutic doses of thalidomide early in pregnancy affects long bone development, leading to foreshortening of the arms or legs, a physical birth defect known as phocomelia. Although very high concentrations of alcohol early in pregnancy can result in similar limb anomalies, alcohol predominately causes "neurodevelopmental" or functional birth defects due to its effects on the nervous system and brain.

The German drug company *Chemie Grünenthal* (now Grünenthal) developed the drug in 1953, and it was first marketed in West Germany on October 1, 1957 under the trade-name Contergan and was licensed in the United Kingdom in 1958. Mainly used as a sedative or hypnotic, thalidomide was used to treat insomnia, anxiety, tension, and gastritis. Doctors prescribed it for pregnant women to prevent nausea and morning sickness. After the success of thalidomide in Europe, the drug company William S. Merrell petitioned the Food and Drug Administration in 1960 to be allowed to market the drug in the United States. However, it never reached the U.S, market because an astute drug reviewer at the Food and Drug Administration (FDA), Dr. Frances O. Kelsey, MD, PhD refused to allow approval of the U.S. licensure.

A lack of evidence that the drug was safe for human use troubled Dr. Kelsey, who was assigned to review the application her first month on the job. The safety studies leading to its European approval had included only rodents, such as rats and mice. Dr. Kelsey's insisting on further research due to insufficient safety data in primates kept thalidomide off

---

[22] *Thalidomide Victims of Canada,* http://www2.awinc.com/users/dascheid/tvac/index.html.
[23] Sadler TW, *Langman's Medical Embryology,* 6th Edition, Williams and Wilkins, Baltimore, 1990, p. 149.

the U.S. market for over a year while drug studies were conducted in pregnant monkeys. In 1961, the Australian physician, William McBride, published a letter in the British medical journal, *Lancet,* after noticing an increase in the number of babies born without limbs to mothers who had taken thalidomide. Soon after, Dr. Kelsey had the results of the thalidomide studies in monkeys demonstrating exactly the same defects. The drug was withdrawn from worldwide markets later the same year.

Over the year the drug was sold, between 5,000 and 7,000 infants were born in Germany with foreshortened limbs (called "phocomelia"), with only 40% of those children surviving into adulthood. Other birth defects involved the heart, eyes, gastrointestinal and urinary tracts, blindness and deafness. Unfortunately, morning sickness, for which thalidomide was most widely used, begins in the first trimester – during the process of "organogenesis," or organ development. This is during a time of rapid changes in the developing embryo, including the formation of long bones in the arms and legs. It turns out that thalidomide disrupts the process of long bone development as well as other organ systems in humans and other primates, but not necessarily in rodents. When taken during early stages of pregnancy for nausea, the women were unknowingly exposing their babies to a potent "teratogen" (chemical that causes birth defects). Prenatal thalidomide exposure caused grotesque, horrific limb defects in which the arms or legs were either undeveloped or presented themselves as stumps or flipper-like protrusions.

The magnitude of the public health crisis that ensued from the thalidomide disaster is more evident by calculating the incidence rate in the population (number of new cases per population) at that time. Throughout the world, about 10,000 total cases were reported of infants with serious problems caused by thalidomide, with only a 50% survival rate worldwide. With the entire world population in 1960 at just over 3 billion people, and in West Germany – around 55 million, over the few years that thalidomide was marketed, there were 10,000 new cases worldwide (or 1 new case per 300,000 people), and in West Germany – 6,000 new cases (0.1 new cases per 1,000). The birth defects caused by prenatal thalidomide exposure led to the drug being banned worldwide for rates much less than what would be considered epidemic proportion.

With a greater number of children having brain damage resulting from maternal alcohol use, why have we not rallied our societal efforts to prevent this chronic epidemic? Based on the proportion of the population with ND-PAE (known as "prevalence"), ND-PAE would qualify as an epidemic. With a U.S. population of 312.9 million in 2012 according the US Census Bureau, the current estimates of 1-5 per 1,000 of school age American children with ND-PAE far exceed the prevalence required to qualify as an epidemic. Yet why is no one calling it that? My educated guess is that, if alcohol caused obvious physical malformations as drastic as thalidomide, that even the alcohol industry would no longer be able to minimize its effects on our offspring. In that way, our society and its children would only be so lucky if alcohol caused such obvious physical birth defects as thalidomide-induced problems. Philosophically, the horrific, graphic damage caused by thalidomide led to society recognizing that its benefits were severely outweighed by the risks of harm to the population. If the damage caused by alcohol was as visible to the naked eye, then it would not have caused so many generations of intellectual issues and disorders for so many people. Surely we would have prevented this from happening as we did by withdrawing thalidomide from world markets and improving pharmaceutical regulations.

> *Although thalidomide was never licensed in the United States, it was distributed as samples to American doctors to try with their patients. It was common practice at that time for drug companies to pass on experimental drugs to doctors, who were then paid to collect data on their patients' results. Patients did not normally know or consent to their part in this loosely controlled research. Dr. Kelsey went on to write the federal rules that continue to govern every clinical drug trial in the United States and was the first official to oversee them.*[24]

---

[24] *"For her part in preventing the distribution of thalidomide in the United States, Dr. Kelsey was awarded the distinguished Federal Civilian Service Medal by President Kennedy on August 7, 1962. Dr. Kelsey retired from the F.D.A. in 2005 at age 90. In September 2010, the F.D.A awarded 96-year-old Dr. Kelsey, the first Kelsey Award in honor of her service. The Kelsey Award will be awarded annually to an F.D.A. staff member."* http://guides.main.library.emory.edu/c.php?g=50422&p=325039

In 1968, the UK manufacturer Distillers Biochemicals, Ltd. (now Diageo) settled a large law suit after a legal battle with the families of those affected. A media campaign in the UK developed to help garner further compensation to victims and their families included a front-page article in the *Sunday Times* - "Our thalidomide children, a cause for national shame." Diageo eventually paid out a total of £28m during the 1970s. This would be approximately $11 million in equivalent 1970s US currency, and about $61 million in 2015 US dollars.

The positive outcome of the thalidomide disaster was development of more structured drug regulations and control over drug use and development. The thalidomide tragedy moved Congress to pass legislation to protect patients from medical experimentation without their consent and to require testing of new drugs before their distribution. Since the time of thalidomide, oversight and monitoring by the Food and Drug Administration requires pharmaceutical companies to accept responsibility for damages caused by their products. Prescription medications and over the counter products are strictly regulated to reduce the risk-benefit ratio such that medical benefits must largely outweigh potential human harm. I can't help but wonder how much better off we'd be if alcohol caused such obvious physical birth defects instead of the equally devastating but more hidden functional birth defects.

Today, thalidomide is being marketed by Celgene as an immunomodulatory drug under the brand names Inmunoprin, Talidex, Talizer, Thalomid, mainly in third world and developing countries (e.g., Brazil) to treat rare conditions, such as leprosy and multiple myeloma (a type of cancer), as well as lupus, leprosy, and as an adjunct for HIV. In order to prevent inadvertent prenatal exposure to this potent teratogen that caused the worldwide birth defect epidemic in the mid-20th century, physicians are ethically and morally bound to advise patients to use two forms of reliable contraception – such as condoms and oral birth control pills or a diaphragm.

So how does ND-PAE compare with Thalidomide rates?

In order to determine the numbers of children affected by prenatal alcohol exposure, the government uses techniques in epidemiology such as surveillance and cross sectional studies, similar to methods used to determine population rates for other neurodevelopmental conditions. Since clinical trials testing alcohol's effects during human pregnancy would be unethical on many levels, epidemiologic studies have necessarily become the gold standard to evaluate the effects of prenatal alcohol exposure in humans. In the information sheet, *FASD by the Numbers: What You Need to Know,* the Substance Abuse and Mental Health Services Administration's FAS Center for Excellence reports that:

> *FASD is the leading known cause of mental retardation. In the United States: Prevalence of FAS in the United States is estimated to be between 0.5 and 2 per 1,000 births. Prevalence of [all FASD] combined is at least 10 per 1,000, or 1 percent of all births. Based on estimated rates of FASD per live births, FASD affects nearly 40,000 newborns each year. The cost to the nation of FAS alone may be up to $6 billion each year. For one individual with FAS, the lifetime cost is at least $2 million.*

According to the Centers for Disease Control and Prevention, the following comparison to other issues raising public health concern points to ND-PAE as a leading preventable epidemic of brain damage:

> ➢ <u>Neurodevelopmental Disorder associated with Prenatal Alcohol Exposure</u> (ND-PAE): *"CDC studies have shown that 0.2 to 1.5 cases of fetal alcohol syndrome (FAS) occur for every 1,000 live births in certain areas of the United States. Other studies using different methods have estimated the rate of FAS at 0.5 to 2.0 cases per 1,000 live births. Scientists believe that there are at least three times as many cases of FASD* [non-dysmorphic ND-PAE] *as FAS."*[25] Studies by Phillip May and colleagues find

---

[25] http://www.cdc.gov/ncbddd/fasd/data.html

that 2-5% (1 in 20) of elementary school children in suburban America have this preventable condition.[26]

➤ Autism: *About 1 in 88 children (11.3 per 1,000 in 2008) has been identified with an autism spectrum disorder (ASD), according to estimates from CDC's Autism and Developmental Disabilities Monitoring (ADDM) Network."*[27]

➤ Cerebral Palsy (CP): *prevalence varied by site, ranging from 2.9 per 1,000 8-year-olds in Wisconsin to 3.8 per 1,000 8-year-olds in Georgia. The average prevalence of CP across the four sites was approximately 3.3 per 1,000 or 1 in 303 8-year-old children in the United States.*[28]

➤ Down's Syndrome: *each year about 6,000 babies in the United States are born with Down syndrome. In other words, about 1 of every 691 babies born in the United States each year is born with Down syndrome.*[29]

➤ Spina Bifida: *...each year, about 1,500 babies are born with spina bifida.*[30]

In other words, more American babies are born each year with ND-PAE than the annual numbers of new cases of autism, cerebral palsy, Down's syndrome, and spina bifida combined with the entire worldwide cohort of infants born with thalidomide embryopathy. Based on an average number of children born each year with ND-PAE (40,000), there would be approximately 1,680,000 individuals with ND-PAE born since 1973 when Smith and Jones wrote their seminal paper characterizing the Fetal Alcohol Syndrome. With costs to the government approximately $6 billion annually, over 42 years this would have been a savings of $252 billion minimum to the U.S. government for cases of FAS (dysmorphic ND-PAE).

---

[26] May PA, Baete A, Russo J, Elliott AJ, Blankenship J, Kalberg WO, Buckley D, Brooks M, Hasken J, Abdul-Rahman O, Adam MP, Robinson LK, Manning M, and Hoyme HE. Prevalence and Characteristics of Fetal Alcohol Spectrum Disorders *Pediatrics*, October 27, 2014.
[27] http://www.cdc.gov/ncbddd/autism/data.html
[28] (http://www.cdc.gov/ncbddd/cp/data.html)
[29] (http://www.cdc.gov/ncbddd/birthdefects/downsyndrome.html)
[30] (http://www.cdc.gov/ncbddd/spinabifida/data.html)

Over the past 50 years, a number of descriptive terms have disguised the brain damage associated with prenatal alcohol exposure amongst other neurodevelopmental disorders. "Minimal brain dysfunction" became widely known as Attention Deficit Hyperactivity Disorder (ADHD), with a fairly large percentage of ND-PAE cases hidden within both conditions. Neurologists use "encephalopathy" to describe brain damage, ranging from static (i.e., not causing changes on encephalography, or EEG) to epileptiform (i.e., causing waveform changes on EEG) and epileptic (i.e., causing overt seizure activity with waveform changes). The resulting "brain irritability" can be seen in 24-hour continuous EEGs as epileptiform activity and spike and wave forms for some individuals. Psychiatrists describe various other neurodevelopmental conditions with terms like autism, pervasive developmental disorder, Asperger's Disorder, learning disabilities, and even schizophrenia. Other neuropsychiatric conditions, such as intermittent explosive disorder (IED), bipolar disorder, and even certain personality disorders camouflage ND-PAE. The quintessential masquerader – ND-PAE is a chameleon condition that shape-shifts and morphs into whatever diagnostic paradigm a clinician chooses to see. The awkward social skills, language deficits, and motor problems look similar to autism or pervasive developmental disorder. The rage episodes, irritability, and aggression masquerades as bipolar disorder or IED. And the odd behaviors, alexithymia, and perceptual deficits may mimic Cluster A or B personality traits.

Don't Blame the Mother!

Often society sees the mother as wholly responsible for a child being born with ND-PAE. A colleague of mine, Laura Riley, JD – an attorney in California, aptly describes the paradox in placing blame on pregnant women:

> *Many have been poorly informed [by the alcohol industry] or may have medical/psychiatric conditions that inhibit their ability to respond to the same messages as social drinkers (i.e., alcohol use disorders). Most people who walk by a visibly pregnant woman smoking would probably cringe.*

> *People might have the same reaction to a pregnant woman who drinks a bottle of whiskey per day. But what about the pregnant woman who is drinking a glass of wine at a cocktail party? Or what about a young woman who is blacked out from a binge episode and having unprotected sex? What about a sexually active alcohol consumer whom has a mistimed pregnancy while adhering to the American Heart Association's (AHA's) advice that daily alcohol use is beneficial? All of these situations are potentially harmful for the fetus. Given the information disparity and mixed messages, it is not surprising that nearly 20% of pregnant women have exposed their babies to alcohol at some stage of pregnancy and nearly one in every eight women of childbearing age drink enough alcohol to affect their offspring during early stages of an unintended pregnancy. Women of childbearing age do not have ready access to all of the information necessary to make decisions relating to ND-PAE. They cannot know what they do not know. Unlike our obligations as citizens to "know" the law insofar as we must comply with it, it is not our obligation to be up to date on medical studies.*

Since the early 1980s, we have known that the major malformations associated with alcohol occur with as little as 4 to 5 servings of alcohol (a binge episode) during the late third to early fourth week post conception (Sulik, 1983). Most women who are social drinkers may not realize that alcohol consumed during episodes of binge drinking at the critical early stages of development may lead to fetal damage. Since this is before most women know they are pregnant, even if a woman stops using alcohol when she finds out she is pregnant, the range of cognitive and physical disability may have occurred already. More about this can be found in Chapter 8.

Figure 4. Critical stages of early development. Our offspring are vulnerable to our lifestyle before we know they exist.

Most Vulnerable Period of Embryo Development

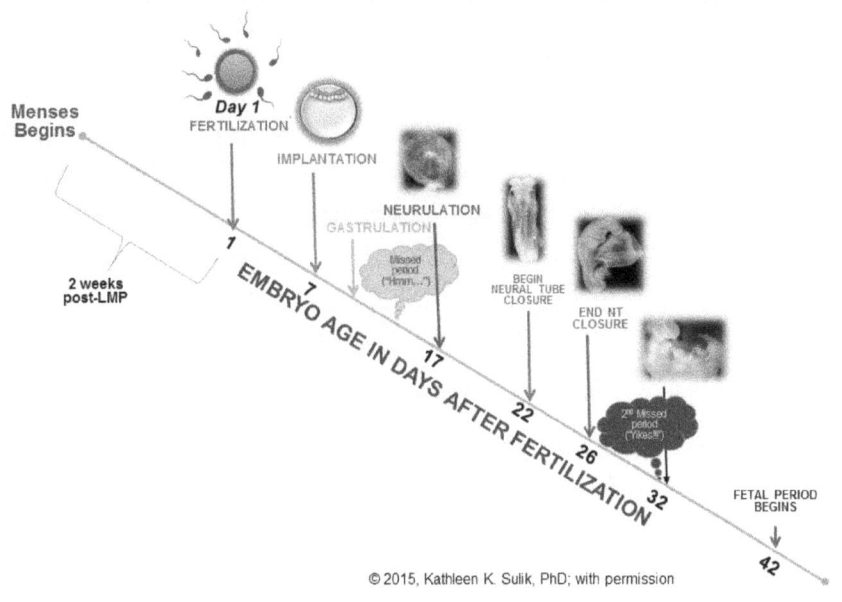

© 2015, Kathleen K. Sulik, PhD; with permission

The current problem is two-fold: approximately 50% of U.S. pregnancies are unplanned (mistimed or unwanted) and nearly 14 million U.S. women binge drink. Detail about unintended pregnancy rates in the U.S. comes from a 2011 article in the journal Contraception. Following a considerable decline from 59 unintended pregnancies per 1,000 women aged 15–44 in 1981 to 49 per 1,000 in 1994, the overall U.S. unintended pregnancy rate has remained essentially flat since then (see chart). Excluding miscarriages, 49% of the pregnancies concluding in 1994 were unintended; 54% of these ended in abortion. Forty-eight percent of women aged 15-44 in 1994 had had at least one unplanned pregnancy sometime in their lives; 28% had had one or more unplanned births, 30% had had one or more abortions and 11% had had both. The proportion of pregnancies that were unintended remained essentially stable between 2001 (48%) and 2006 (49%). (Finer, et al., 2011).

According to the Centers for Disease Control and Prevention (CDC), a recent CDC Morbidity and Mortality Weekly Report (MMWR) recognized binge drinking as a "serious, under-recognized problem among women and girls," with nearly 14 million US women binge drinking. Women who binge drink do so frequently – about 3 times a month – and have about 6 drinks per binge. [Binge drinking for women is defined as consuming more than three alcohol drinks (beer, wine, or liquor) on an occasion (i.e., over the course of an evening at home or socializing with others)]. Even more alarming is the age disparity in the drinking patterns. While 1 in 8 women aged 18 years and older report binge drinking, 1 in 5 high school girls binge drink. (MMWR, 2013) Attorney Laura Riley believes that it is impossible to charge a mother for liability for inadvertently exposing her child to alcohol prior to pregnancy recognition.

> *There is a difference between moral responsibility and legal liability. In order to be 'criminally liable,' intent is necessary—and it must be shown that the intent that is tied to the action taken (this is the legal doctrine of 'concurrence' in the U.S.). Black's Law Dictionary defines intent as 'The state of mind accompanying an act, especially a forbidden act ... intent is the mental resolution or determination to do (some act)' (Reuters, 2009). That is, it would be necessary to show that there was at least knowledge that, by consuming alcohol, she was harming the fetus in the way that it was harmed (suffering from ND-PAE). Being that most women do not know they are pregnant until they are about two months into their pregnancy, this intent is paradoxically impossible to impose for many women. For those who have a psychiatric condition such as alcohol use disorder, better screening and treatment programs are needed for reproductive age women before they become pregnant, as well as widespread contraceptive use for all alcohol consumers.*
>
> *Another form of liability is tort liability, which can remove the intent requirement. Two modes of tort liability are strict liability and negligence. Strict liability is typically reserved for actions that are so rash and destructive that they should be punished even if intent would be difficult to prove–like owning wild animals that harm others, or statutory rape.*

> *Negligence can arise where someone participates in the creation of risk, the risk that creates a situation resulting in harm to another. In order for there to be negligence liability, there must be a duty owed—in this case by the mother to the fetus. Without getting in to more controversial issues, courts have not recognized that mothers owe duties regarding alcohol consumption to an unborn child. And rightly so, since frequently alcohol consumption by pregnant women is a result of alcoholism or a lack of awareness that she is even pregnant. There is, however, a duty owed by health care professionals to patients, including alcohol-consuming female patients of childbearing age who may be at risk of unintentional pregnancy, therefore a binge episode before pregnancy recognition.*

In short, the government, physicians, and the alcohol industry can do more to adequately inform women of this little known fact. The only public health warning related to the harm that alcohol may cause a developing child is an industry agreed upon label that currently reads: "According to the Surgeon General, women should not use alcoholic beverages during pregnancy because of the risk of birth defects." The reproductive health effects of alcohol use exist on a continuum, not simply causing "a risk of birth defects."

It is ironic that our society places more emphasis on protection of our food supply and medications than on the brains of its future citizens. The pharmaceutical industry is so highly regulated that medications must undergo years of testing to prove that they are both safe and effective for treatment of illnesses, and that their benefits outweigh their risks to society. The food industry is regulated to the degree that even the chemical contents are scrutinized by government oversight, yet alcohol which we know is harmful to the developing fetus is relatively unregulated with regard to labeling. In a book edited by Dr. Monty Nelson, Laura Riley and I make the point that – unlike the food and drug industries, the alcohol industry has no Federal mandate for prevention of this widespread public health crisis due to its products.[31] It has not been held accountable for the permanent brain damage to children due to use

---

[31] Rich SD and Riley LJ. Neurodevelopmental Disorder Associated with Prenatal Alcohol Exposure: Consumer Protection and the Industry's Duty to Warn. Chapter 3 in *Fetal Alcohol Spectrum*

during early stages of pregnancy – statistically the highest number of children affected since most women are unaware they are pregnant and may inadvertently expose their offspring during that vulnerable time. A later chapter on *Preconceptional Perspectives* reveals solutions to the moral and ethical paradox of binge drinking and unintended pregnancy.

In her book, *Fetal Alcohol Spectrum Disorders: Trying Differently Rather than Harder,* Diane Malbin discusses the never ending list of unhelpful and sometimes detrimental diagnoses that individuals with ND-PAE are inevitably labeled with: Attention Deficit Hyperactivity Disorder, Learning Disabilities, Antisocial Personality Disorder, Speech/Language Disorder, various disorders on the Autistic Spectrum, and Conduct Disordered to name a few. Other misdiagnoses for these patients include Bipolar Disorder, Oppositional Defiant Disorder, Schizotypal and Schizoid Personality Disorder, Borderline Personality Disorder, and Schizophrenia. A 2014 study published in the journal Psychiatric Services in Advance by Dr. Carl Bell found that nearly 40% of 611 adult and adolescent psychiatric patients who attended the Family Medicine Clinic at Jackson Park Hospital on Chicago's Southside had clinical profiles consistent with ND-PAE. In comparison, only around 9% of the patients had true DSM-5 neurodevelopmental disorders, such as intellectual disability, ADHD, or autism. Others had been misdiagnosed with mental illness such as bipolar disorder. Like most psychiatrists, Bell, who is a staff psychiatrist at the Family Medicine Clinic, admits missing such cases in the past. He is quoted by *Psychiatric News* that he " …'did not have a clue how prevalent this neurodevelopmental problem was, and … was seeing these people clinically.' Bell believes this issue is … the biggest public-health issue since polio.'"

Most psychiatric disorders are thought to have a neurodevelopmental component, with genetic, prenatal, and postnatal influences on their severity, trajectory, and prognosis. One such condition, schizophrenia, has been described as a "developmental encephalopathy" with neurodevelopmental causes and prodromal symptoms that appear

---

Disorders in Adults: Ethical and Legal Perspectives - An overview on FASD for professionals. Nelson M and Trussler M (Eds.). Int. Library Ethics, Law Volume Number:63. Springer Publications, 2015.

autistic-like (i.e., social communication issues, intellectual deficits, and soft neurological signs). Psychiatric researchers like E. Fuller Torrey have pointed to biological causes of schizophrenia, such as viral infection in the second trimester. Mary Greene in her eloquent memoir, *When Rain Hurts,* about her adopted son Peter who has ND-PAE and autism, describes his auditory and visual hallucinations leading to him being on several medications, including two highly potent atypical antipsychotics. In my clinical experience, many patients with ND-PAE develop into adolescents and adults with increasingly suspicious and paranoid-like behaviors due to problems with social perception and anxiety. Some have anosmia due to a deficient olfactory bulb – a midline nerve in the brain (cranial nerve 1) responsible for the sense of smell, leading to deficits in perception of their own body odor that interfere with their hygiene.

Does the profit from alcohol sales outweigh the social consequences of ND-PAE?

Western society is infatuated with alcohol's inebriating qualities, policies are influenced by the strong lobby of the alcohol industry, and our governments benefit from taxes on alcohol importation and sales. According to one distilled spirits industry website, they contribute $110 billion to the American economy each year, employing more than 1.2 million American workers in sales, production and distribution of alcoholic beverages.[32] The Brewers Association boasted an annual revenue of $101 billion in 2014, with craft brewers contributing $33.9 billion to the U.S. economy in 2012, more than 360,000 jobs.[33] The Wine Institute reports a staggering 2.82 gallons of wine consumed per resident in 2013, with a total of 892 million gallons sold in the U.S. annually – at a total of $16.7 billion spent on wine consumed at home and at full-service restaurants.[34]

In Montgomery County, Maryland annual revenue generated from the sales of alcohol averages about $35 million ($5 million more than our

---

[32] The Distilled Spirits Council of the United States. http://www.discus.org/about/industry/
[33] The Brewers Association. https://www.brewersassociation.org/statistics/national-beer-sales-production-data/
[34] *Statistica*, The Statistics Portal. http://www.statista.com/topics/1541/wine-market/

national research budget on FASD), yet not one sign is posted warning childbearing age women about the potential for harm to their offspring. Up to 41 other states now require point of purchase labeling thanks to efforts of the Center for Science in the Public Interest and the National Organization on Fetal Alcohol Syndrome.[35] Yet such efforts fall short of targeted advertising by the alcohol industry of new "girlie" beer, malt liquor, and wine to childbearing age women - contributing to a national public health catastrophe with international repercussions. Nationwide, one in eight high school girls consumes at least 5 drinks of alcohol at least once per week and one in five 18 to 44-year-old women consume a binge amount as often as once per week. With beer and wine consumed *en masse* as a social lubricant, and Western young women now drinking more hard liquor than their male counterparts, countless babies have already been exposed to moderate to heavy amounts of alcohol in the womb before we know they exist. These high rates of binge drinking in the face of a 50% U.S. unplanned pregnancy rate (i.e., alcohol use and unprotected sex) has led to a hidden but preventable epidemic of brain damage in our future citizens.

Considering the revenue generated by the alcohol industry in excess of $227 billion annually, a reasonable educated guess for the amount spent on preventing the brain damage associated with prenatal alcohol exposure may be at least $1 billion. However, the government's funding priorities on the revenue generated by the $23 billion tax on alcohol, tobacco and fire arms does not even include a line item for ND-PAE. The mission statement of the Alcohol and Tobacco Tax Trade Bureau includes consumer protection involving labeling compliance:

> *To collect the Federal excise taxes on alcohol, tobacco, firearms, and ammunition, and assure compliance with Federal tobacco permitting and alcohol permitting, labeling, and marketing requirements to protect consumers.*[36]

---

[35] State Action Guide: Mandatory Point-of-Purchase Messaging on Alcohol and Pregnancy. Center for Science in the Public Interest. October 2008 https://www.cspinet.org/new/pdf/state_action_guide.pdf

[36] *President's Budget*. Alcohol and Tobacco Tax and Trade Bureau, p. 3, FY 2014.

However, funding priorities lack strategic direction for consumer protection involving warnings to reproductive age individuals about consumption of alcohol prior to pregnancy recognition (or even during pregnancy):

FY 2014 Priorities[37]

- Collect roughly $23 billion in annual excise tax revenues due to the Federal government;
- Complete audits and investigations of TTB taxpayers based upon risk and random selection to ensure lawful operations and tax reporting and payment compliance;
- Detect and address criminal diversion activity, including counterfeit alcohol and tobacco products, to protect the Federal revenue stream, U.S. consumers, and fair market activity;
- Reduce the burden of compliance by promoting electronic filing options for industry members, including the online system that allows industry members to apply for an original permit to start a new alcohol or tobacco business;
- Conduct statistically valid marketplace sampling programs to determine industry-wide compliance with Federal regulations for alcohol beverages, and tailor enforcement and education programs based on these findings to ensure products sold to U.S. consumers meet
- Federal alcohol beverage production, labeling, and marketing requirements;
- Refine TTB product safety activities that focus on the integrity and safety of domestic and imported alcohol beverage products;
- Promote U.S. exports by facilitating industry compliance with foreign requirements and by working with foreign regulators to reduce barriers to trade; and
- Strengthen global tax administration structures through work with trading partners and emerging markets to prevent tax loss from illicit trade and improve tax collection outcomes.

---

[37] *President's Budget.* Alcohol and Tobacco Tax and Trade Bureau, p. 4, FY 2014.

In contrast, one government funded study estimated in 2003 that annual costs for FAS (accounting for only 20% of all FASD cases) are a staggering $5.4 billion, which would be more than $26 billion for all cases of ND-PAE. These figures do not account for the costs of legal, criminal justice, and other secondary outcomes for affected individuals. Does the annual financial gain of $222 billion justify the costs to the future intellectual potential of our growing population? Are we doing enough to prevent and ameliorate the problem?

Many women with unintended pregnancies who are using alcohol regularly or in binge amounts may inadvertently expose their babies to alcohol, leading to ND-PAE before they even know they are pregnant. The obscure "warning label," pointing to possible birth defects as a reason to avoid alcohol during pregnancy, is inadequate to fulfill even a minimal duty to warn. Might some of the alcohol revenue go into more accurate warnings on its products to protect the consumer from risk? What about alcohol manufacturers being required to distribute condoms where alcohol is sold in order to be a bolder, more definitive warning about the dangers of alcohol use and unintended pregnancy?

More widespread prevention and treatment initiatives are needed to educate people about the damage of alcohol use prior to pregnancy recognition, identify and treat women with alcohol and substance abuse issues prior to pregnancy, and provide greater access contraception and preconception health programs. Women who do become pregnant should be supported to maintain a substance free lifestyle before, not just during, pregnancy and provided assistance for parenting their children rather than having multiple substance exposed babies because they have been taken away by the system.

Beginning life with mild to moderate brain damage and developing resultant psychiatric illnesses from the psychological trauma they grow up with, individuals with ND-PAE require reasonable job skills for gainful employment. Ignoring this financially and intellectually impairing public health crisis perpetuates a centuries old human rights tragedy for individuals, families, and communities – interwoven with school failure, crime, violence, un-employability, decreased lifetime productivity, homelessness and destitution. Lacking the self-esteem and wherewithal to work their way into the working poor or working class, they fall into the chasm of the underclass – relegated to the fringe of society's disenfranchised.

The upcoming chapter on *The American Social Caste System*, highlights the perfect storm creating Huxley's modern day *Brave New World* in our own oblivious society.

Chapter 5

THE AMERICAN SOCIAL CASTE SYSTEM

I understood the problems plaguing poor communities of color, including problems associated with crime and rising incarceration rates, to be a function of poverty and lack of access to quality education—the continuing legacy of slavery and Jim Crow. Never did I seriously consider the possibility that a new racial caste system was operating in this country. The new system had been developed and implemented swiftly, and it was largely invisible, even to people, like me, who spent most of their waking hours fighting for justice.

— Michelle Alexander, The New Jim Crow,
The New Press, New York. 2010.

Twenty-first century headlines bleed accounts of socially-disenfranchised African American youth with varying degrees of functional birth defects limiting their capacity for enlightenment. Birthed into the underclass, their brain damage acquired in the womb seals their fate – relegating them to the impoverished, incarcerated, underserved, homeless and marginalized citizens who rely on public assistance and government subsidies to survive in a highly technical society where the smartest, intellectually savvy thrive. Sociologists estimate that the American society is broken into several classes depicted in in Figure 5 – a rendition of the modern caste system proposed by Dennis Gilbert. His political and social context is similar to the caste system in pre-colonial India and ancient Egypt. In his 1998 book, Gilbert shows a consistent pattern of growing inequality since the early 1970s based on distribution of earnings and residential segregation by class. He cites nine key variables influencing class designation: occupation, income, wealth, prestige, association, socialization, class consciousness, power, and social mobility.[38]

Figure 5: Our *Brave New World* – U.S. Social Stratification[39]

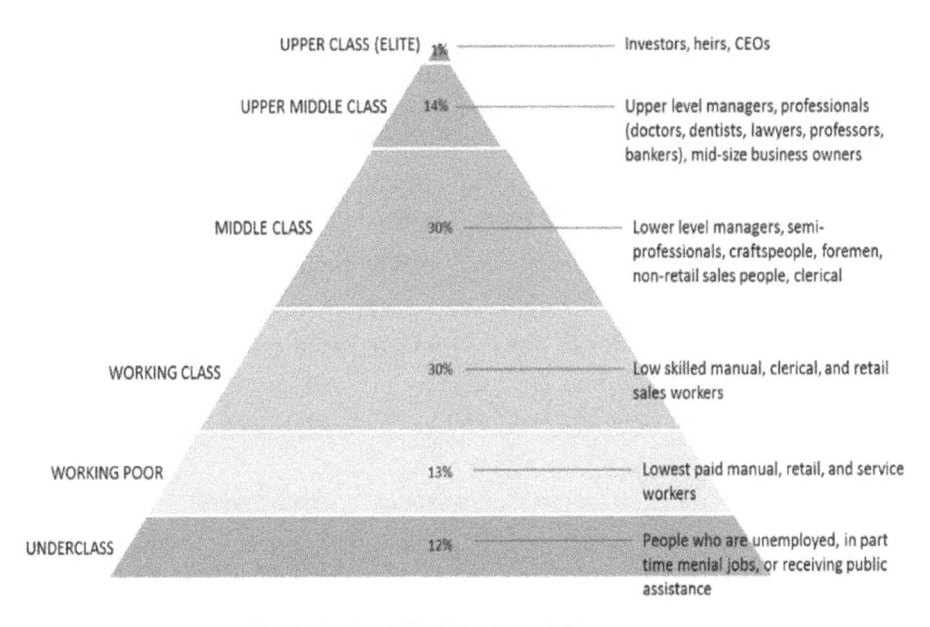

```
UPPER CLASS (ELITE)   1%  ──── Investors, heirs, CEOs
UPPER MIDDLE CLASS   14%  ──── Upper level managers, professionals
                                (doctors, dentists, lawyers, professors,
                                bankers), mid-size business owners
MIDDLE CLASS         30%  ──── Lower level managers, semi-
                                professionals, craftspeople, foremen,
                                non-retail sales people, clerical
WORKING CLASS        30%  ──── Low skilled manual, clerical, and retail
                                sales workers
WORKING POOR         13%  ──── Lowest paid manual, retail, and service
                                workers
UNDERCLASS           12%  ──── People who are unemployed, in part
                                time menial jobs, or receiving public
                                assistance
```

© 2016; Susan D. Rich, MD, MPH
Adapted from *The American Class Structure*, Dennis Gilbert, 1998.

In public health, sociology, and medicine, we know that both risk and resiliency factors contribute to a person's ability to rise above their social or family problems – staying out of jail and avoiding substance abuse problems. Unlike the socially disenfranchised "untouchables" in India, many of us have been able to pull ourselves out of the relative "lower classes" – the underclass, working poor, or working class – by an advanced education, perseverance, and/or hard work. Research shows that one's internal resiliency factors contribute more to success,

---

[38] Gilbert D. *The American Class Structure in an Age of Growing Inequality.* Wadsworth Publishing; 6th Edition; September 27, 2002.
[39] Adapted from *The American Class Structure*, Dennis Gilbert, 1998.

self-actualization and upward social mobility than risk factors. Individual protective factors of academic motivation, an eagerness to please, and a Protestant work ethic led to me attending college then graduate school and medical school in order to rise above my family's social status. Determination, drive, and advanced education boosted me into the upper middle class, though still well below the elite caste. Intellect, flexibility, adaptive functions, social skills, emotional regulation, and belief in myself contributed to my success - all attributes frequently stripped from individuals exposed to alcohol in the womb.

The Centers for Disease Control and Prevention explain the impact of various risk and resiliency (protective) factors in the perpetuation of violence by American youth.[40] An understanding of risk (Figure 6) versus resiliency (Figure 7) factors helps appreciate that individuals with ND-PAE have the odds stacked against them from birth. They are both vulnerable to risk of victimization, perpetration, and violence and less resilient to overcome the forces that perpetuate these cycles in their families and communities.

---

[40] *Youth Violence: Risk and Protective Factors.* Injury Prevention & Control; Division of Violence Prevention; Centers for Disease Control and Prevention.

Figure 6. Risk Factors for the Perpetration of Youth Violence. The following risk factors recognized by the CDC as associated with perpetration of youth violence are also common among individuals with ND-PAE:

Individual Risk Factors
["Intrinsic Vulnerabilities"]

- ✓ History of violent victimization
- ✓ History of early aggressive behavior
- ✓ Low IQ
- ✓ Antisocial beliefs and attitudes
- ✓ Attention deficits, hyperactivity or learning disorders
- ✓ Deficits in social cognitive or information-processing abilities
- ✓ History of treatment for emotional problems
- ✓ Exposure to violence and conflict in the family
- ✓ Involvement with drugs, alcohol or tobacco
- ✓ Poor behavioral control
- ✓ High emotional distress

External Risk Factors
["Extrinsic Vulnerabilities"]

Family Risk Factors	
• Harsh, lax or inconsistent disciplinary practices	• Authoritarian childrearing attitudes
• Low emotional attachment to parents or caregivers	• Low parental involvement
• Parental substance abuse or criminality	• Low parental education and income
• Poor monitoring and supervision of children	• Poor family functioning

Peer and Social Risk Factors	
• Association with delinquent peers	• Involvement in gangs
• Social rejection by peers	• Lack of involvement in conventional activities
• Poor academic performance	• Low commitment to school and school failure

Community Risk Factors	
• High concentrations of poor residents	• Diminished economic opportunities
• High level of family disruption	• High level of transiency
• Socially disorganized neighborhoods	• Low levels of community participation

Figure 7. Protective Factors for the Perpetration of Youth Violence. The Centers for Disease Control and Prevention lists the following protective factors based on preliminary studies:

Individual Protective Factors	
• Intolerant attitude toward deviance • High grade point average (as an indicator of high academic achievement) • High IQ	• Positive social orientation • Highly developed social skills or competencies • Religiosity • Highly developed skills for realistic planning
Family Protective Factors	
• Connectedness to family or adults outside the family • Ability to discuss problems with parents • Consistent presence of parent during at least one of the following: when awakening, when arriving home from school, at evening mealtime or going to bed	• Perceived parental expectations about school performance are high • Involvement in social activities • Parental / family use of constructive strategies for coping with problems (provision of models of constructive coping) • Frequent shared activities with parents
Peer and Social Protective Factors	

• Possession of affective relationships with those at school that are strong, close, and pro-socially oriented
• Commitment to school (an investment in school and in doing well at school)
• Close relationships with non-deviant peers
• Membership in peer groups that do not condone antisocial behavior
	• Involvement in prosocial activities
	• Exposure to school climates that characterized by:
	  ✓ Intensive supervision
	  ✓ Clear behavior rules
	  ✓ Consistent negative reinforcement of aggression
	  ✓ Engagement of parents and teachers

The Silent Epidemic: A Child Psychiatrist's Journey beyond Death Row

Youth risk assessment research points to resiliency factors that promote successful outcomes. The Child Welfare Information Gateway defines resiliency as *the ability to cope and adapt to change. Being resilient allows children and youth to overcome difficulties in their lives.*[41] An individual's resiliency or protective factors enable the person to traverse adolescence and young adulthood free from mental health issues, institutionalization, criminality, substance abuse, employment problems, and a myriad of other secondary disabilities. These protective factors reduce their risks of becoming violent yet have not been studied as extensively or rigorously as risk factors.

In her years of research and clinical work, Dr. Anne Streissguth of the University of Washington at Seattle followed a cohort of over 415 individuals with ND-PAE. She used the term "secondary disabilities"[42] to describe the range of social problems and poor outcomes individuals with ND-PAE develop throughout their life course as a result of lack of supports and services. Identifying high rates of mental health issues in her population, she characterized the following six main categories[43] of secondary disabilities based on composite answers to 450 questions asked during life history interviews:

- **Mental Illness:** 94% of the 415 patients in Dr. Streissguth's cohort were diagnosed with at least one mental health condition, including ADHD during childhood (60%); and clinical depression in almost all the patients by adulthood, with high rates of attempted suicide (23%) and suicidal threats (43%).
- **Academic Disruptions:** 43% of school age children and 70% of adults in her sample had either been suspended from, expelled from or dropped out of school. *Common school problems include: not paying attention; incomplete homework; can't get*

---

[41] https://www.childwelfare.gov/topics/systemwide/mentalhealth/common/resiliency/?hasBeenRedirected=1

[42] Secondary disabilities are defined as those issues occurring as a result of ND-PAE and can be prevented or lessened by better understanding and appropriate interventions. Secondary disabilities were ascertained from life history interviews of 415 individuals with FASD using 450 questions.

[43] Streissguth AP, Barr HM, Kogan J & Bookstein FL. *Understanding the Occurrence of Secondary Disabilities in Clients with Fetal Alcohol Syndrome (FAS) and Fetal Alcohol Effects (FAE),* Final Report to the Centers for Disease Control and Prevention (CDC), August, 1996, Seattle: University of Washington, Fetal Alcohol & Drug Unit, Tech. Rep. No. 96-06, 1996.

along with peers; disruptive in class; disobeying school rules; talking back to the teacher; fighting; and truancy.[44]
- **Legal/Criminal Issues:** 42% of Dr. Streissguth's cohort, and 60% of those age 12 and over had been involved with law enforcement, charged or convicted of crime (initially shoplifting). *The most common crimes committed (by almost half of individuals with FASD age 12-20) were crimes against persons (theft, burglary, assault, murder, domestic violence, child molestation, running away), followed by property damage; possession/selling; sexual assault; and vehicular crimes.* [45]
- **Institutionalization:** 60% of those age 12 and over had received inpatient psychiatric hospitalization, alcohol/drug treatment, or incarceration for crime; with over 40% of adults having been incarcerated; 30% confined to a mental institution; and 20% receiving substance abuse treatment.
- **Sexual Misconduct:** 45% of those age 12 and over, and 65% of adult males had problems associated with sexual behaviors, including but not limited to incarceration or treatment. These figures are thought to underestimate the true incidence of sexual misconduct (embarrassment or fear incarceration may prevent some cases from being reported). *Problem sexual behaviors most common with FASD include: sexual advances; sexual touching; promiscuity; exposure; compulsions; voyeurism; masturbation in public; incest; sex with animals; and obscene phone calls.* [46]
- **Alcohol/Drug Problems** were experienced by 30% of individuals age 12 and over. Of the adults with FAE, 53% of males and 70% of females experienced substance abuse problems. This is more than 5 times that of the general population.

Only 8% of the individuals in the study had no problem with independent living or employment. To determine levels of independence in adulthood,

---

[44] Kellerman T. *Secondary Disabilities in FASD*. Published online 2000-02; Revised May 31, 2003.
[45] Kellerman T. *Secondary Disabilities in FASD*. Published online 2000-02; Revised May 31, 2003.
[46] Kellerman T. *Secondary Disabilities in FASD*. Published online 2000-02; Revised May 31, 2003.

two additional categories were identified for individuals 21 years of age and older (median age 26):

- **Dependent Living:** 80% of adults with ND-PAE were not living independently (i.e., were living in group homes or other types of supervised residential settings).
- **Employment Problems:** 80% of adults with ND-PAE had difficulty remaining employed for any reasonable length of time.

Dr. Streissguth identified the greatest risk factors associated with secondary disabilities in ND-PAE to be:

- **IQ over 70** (those with lower IQ's are likely to get more services and intervention). This means that individuals without intellectual disability are at greater risk those with intellectual disability.
- **Exposure to violence** (sexual and/or physical abuse), which occurs at rate of 72% of individuals with ND-PAE. Those exposed to violence are four times as likely to exhibit inappropriate sexual behavior.
- **Problems with Parenting:** Of the 100 females of childbearing age, 30 had given birth; 40% drank during pregnancy, more than half no longer had the child in their care. Of their children, 30% have been diagnosed with or suspected of having ND-PAE. More recent studies have shown that a cycle of prenatal alcohol exposure within families perpetuates ND-PAE for several generations.

Dr. Streissguth identified several protective factors to predict improved outcomes and prognosis for individuals with ND-PAE:

- **Early diagnosis** is a universal protective indicator for all secondary disabilities. Only 11% of individuals with ND-PAE were diagnosed by age 6. She concluded that every effort must be made to attain early diagnoses for children.

- **Eligibility for services** from state programs for developmental disabilities is another strong protective factor. These services are needed by most individuals with ND-PAE, yet most do not qualify based on the strict requirements for intellectual disability.
- **Living in stable home** with nurturing parents and minimum of changes in household. Such stable environments are missing for children in foster care or living with a parent who has ND-PAE who may have limited parenting skills.
- **Protection from violence**, that is not witnessing or being victimized by violence.

Perhaps the gravest "secondary disability" in individuals with ND-PAE identified by Dr. Anne Streissguth, homelessness, is the worst outcome possible for individuals in the underclass – relegated to living at the lowest level of society without the means to provide even the most basic of Maslow's hierarchical needs. Hidden between the lines of news articles and headlines, the truth emerges – veiled in the hangover of last night's drunken stupor. Recently, *Science Daily*, an online newspaper reported that a majority of homeless Canadians with mental illness have cognitive deficiencies, including learning disabilities, memory deficits and problem solving issues.[47] These are all commonly seen effects of prenatal alcohol exposure.

In other words, the proverbial pink elephant sitting in the middle of the dilapidated dining room table – is that alcohol, our socially-sanctioned, blameless intoxicant, is the cause of suffering of many individuals relegated to the underclass. The one assured way to preserve our modern brave new world outlined by Aldous Huxley in 1932 is to expose babies to alcohol in the womb even before pregnancy recognition. By inundating communities with alcohol, failing to require proper consumer protection by accurate labeling, and gaining profit to drive economic growth and development despite the incapacitation of the unborn, we have created a modern day system of slavery, servitude, and compliance. Perhaps one of the greatest thought leaders of our

---

[47] *Science Daily*, January 26, 2015.

time in psychiatry, Dr. Carl Bell – an adult psychiatrist practicing in a community health center in south side Chicago, Illinois, expresses the magnitude of this problem best in a recent newsletter article from the National Association of County Behavioral Health and Developmental Disabilities Directors:

> *...One of the more obvious social determinants of health that has heretofore been overlooked is the plethora of liquor stores in low-income African-American communities. We recognize 'food deserts,' but are less familiar with the concept of 'food swamps,' i.e. communities plagued with an overabundance of harmful provisions. In this case, the liquid of the swamp is alcohol.*
>
> *Since 1968, I have been observing a phenomenon in low-income African-American communities that I did not understand until more recently. This phenomenon is characterized by a disproportionate number of African-Americans with intellectual disability, explosive tempers, and an inability to flourish as evinced by poor school performance, lousy employment outcomes, and intra-psychic and interpersonal difficulties. Individuals with such problems populate special education classes, juvenile and adult correctional facilities, psychiatric institutions, and child protective programs within our Nation's county medical and social service systems. I had observed this phenomenon in 2/3rds to 3/4ths of African-American children in the Cook County Temporary Juvenile Detention Center[48] and since 1977 in special education classes in the Chicago Board of Education.[49] Explanations for these problems ranged from genetic inferiority to what was referred to as 'sociocultural mental retardation' during my time at Meharry Medical College; I never believed either.*
>
> *After 45 years of wondering, it all finally makes sense: These individuals suffered from fetal alcohol exposure resulting in Fetal Alcohol Spectrum Disorders (FASD)/Alcohol Related Neurodevelopmental Disorders (ARND), or what is currently*

---

[48] Bell CC & Jenkins EJ. Violence prevention and intervention in juvenile detention and correctional facilities. *Journal of Correctional Health Care*, Vol. 2, Issue 1: 17-38, 1995.

[49] Bell CC, Preventive Psychiatry in the Board of Education. *Journal of the American Medical Association*; Vol. 71, NO. 9, pp. 881-87; 1979.

> defined in DSM-5 as Neurodevelopmental Disorder Associated with Prenatal Alcohol Exposure. About a year and a half ago, we began to actively observe this constellation of symptoms in patients seen at Jackson Park Hospital's Family Medicine Clinic, a facility serving a predominately African-American population of 143,000 whose median income is $33,809. Of 611 patients (ages 4-78; 95% on Public Assistance for medical care) that were seen, 297/611 (49%) evidenced Neurodevelopmental Disorders dating back to childhood, including 237/611 (39%) who specifically reported symptoms and histories consistent with Neurodevelopmental Disorder Associated with Prenatal Exposure to Alcohol. This is an extraordinarily high prevalence rate.
>
> The vast majority of these fetal alcohol cases arose when an African- American woman, who did not know she was pregnant, drank alcohol, only to realize that she was pregnant and then stop drinking. Unfortunately, the damage was already done. Their children's histories were replete with reports of special education for intellectual disability, ADHD, learning disorders, and behavioral difficulties that were couched in affect dysregulation – explosive, unpredictable tempers. Histories of incarceration in juvenile and adult corrections and long-standing unemployment were also common.[50],[51]

Dr. Bell's enlightenment to the social disparities brought on by prenatal alcohol exposure was possible because of his training and experience in science, epidemiology, and social science. In this way, education combined with one's perspective can lift the veil of disillusionment about a human rights problem for a future generation. In her groundbreaking book: *The New Jim Crow: Mass Incarceration in the Age of Colorblindness,* Michelle Alexander takes on a single but important aspect of the issue – the overwhelming racial disparity in incarceration rates. Just as abolitionists propelled the ending of slavery as

---

[50] Bell CC. Prevalence of Fetal Alcohol Exposure in Low-Income African Americans. *Newsletter of the National Association of County Behavioral Health and Developmental Disabilities Directors.* January 23, 2015.

[51] Bell CC and Chimata R. Prevalence of Neurodevelopmental Disorders Among Low-Income African Americans at a Clinic on Chicago's South Side. *Journal of Psychiatric Services;* 66(5), pp. 539–542; May 1, 2015.

a societal norm in the latter 19th century, *The New Jim Crow* has spawned an awakening in social consciousness of this twenty first century human rights issue. A missing link from her comprehensive look at the issue is the impact of prenatal alcohol exposure on reduced vocational potential, increased susceptibility for antisocial behaviors, and higher rates of incarceration for young adults with ND-PAE, a condition with a higher prevalence rate in minority communities.

Although our great country celebrated its first African American President in the twenty first century, an increasing percentage of black men are relegated to the underclass by being incarcerated in our corporate correctional system, further enslaved by lifelong criminal records. Professor Alexander points out the warehousing of African American men in jails, prisons, and parole systems as denying their basic human and civil rights to vote, serve on juries, hold a legitimate job, have fair housing opportunities, and be provided access to public education and other benefits. In her eloquent manuscript, the civil-rights-lawyer-turned-legal-scholar equates discrimination against those individuals who have been incarcerated with policies in the 19th century discriminating against African Americans, known by the term "Jim Crow" laws. Since felons are denied basic privileges and inalienable rights afforded to other citizens, this "catch 22" allows arcane discrimination practices to be socially sanctioned. In her words, *we have not ended racial caste in America; we have merely redesigned it.*[52] Her point is well made, that the War on Drugs particularly targets African American men, further destabilizing their families and communities. Her view is that the U.S. criminal justice system has become a profitable form of racial control, even while espousing colorblindness.

The hidden impact of prenatal alcohol exposure on race, poverty, and social class is again found embedded between the lines of a recent article[53] by Terrance McCoy in the *Washington Post*. Freddie Gray, the subject of the article, lost his life in Baltimore, Maryland as a result of police negligence that led to several days of rioting, reminiscent of

---

[52] Alexander M. *The New Jim Crow*. The New Press, New York, 2010. http://TheNewJimCrow.com/about.
[53] McCoy T. Freddie Gray's life: a study in the sad effects of lead paint on poor blacks. *Washington Post*, April 29, 2015.

the 1960s civil rights demonstrations. McCoy briefly mentions that Gray's mother used heroin and he was born premature. While lead exposure was highlighted in the article, no mention was made about the possibility of brain damage associated with prenatal alcohol exposure. Notwithstanding the harmful effects of prenatal and postnatal lead exposure, the neurocognitive, mood dysregulation ("mood swings" and aggression), and other neuropsychiatric issues were likely due to ND-PAE, not simply lead exposure.

> *Freddie Gray's path toward such litigation began months after his birth in August of 1989. He and his twin sister, Fredericka, were born two months prematurely to a mother, Gloria Darden, who said in a deposition she began using heroin when she was 23. He lived in the hospital his first months of life until he gained five pounds. It wasn't long after that he was given the first of many blood tests, court records show. The test came in May of 1990, when the family was living in a home on Fulton Avenue in West Baltimore. Even at such a young age, his blood contained more than 10 micrograms of lead per deciliter of blood — double the level in which the Center for Disease Control urges additional testing. Three months later, his blood had nearly 30 micrograms. And then, in June of 1991 when Gray was 22 months old, his blood carried 37 micrograms.*

While early lead exposure can cause a range of neurodevelopmental issues, including cognitive impairment and attention deficits, studies have found a "neurobehavioral signature" limited to executive functions (attention, processing speed, working memory), intellect, reaction time, fine motor skills, visual–motor integration, and teacher-reported withdrawn behaviors.[54] Increased exposure is also associated with neuropsychiatric disorders such as attention deficit hyperactivity disorder and antisocial behavior.[55]

To my knowledge, there is limited evidence to support lead exposure as an antecedent to violent, antisocial behavior while prenatal alcohol

---

[54] Chiodo LM, Jacobson SW, Jacobson JL. Neurodevelopmental effects of postnatal lead exposure at very low levels. *Neurotoxicology and Teratology* 2004, 26:359 – 371.

[55] Bellinger DC. Very low lead exposures and children's neurodevelopment. *Current Opinion in Pediatrics.* 2008, 20:172–177.

has been well studied as causing faulty neuronal wiring leading to mood dysregulation and autonomic arousal that Gray suffered from according to reports. Children with both ND-PAE and post-natal lead exposure (e.g., those adopted from foster care, from Eastern European countries, as well as from biological families) have a greater probability of neurodevelopmental deficits than with either exposure alone. As would be expected, their neurodevelopmental outcomes are much worse than those with prenatal alcohol exposure alone, or prenatal alcohol exposure and neglect or abuse.

Even if Gray was privileged enough to be born into or adopted into the elite – an heir to a fortune or son of a fortune 500 CEO, he would have less chance of being an "alpha" in Huxley's *Brave New World* than becoming President of the United States. Like others in the 12% of society making up the underclass – either in menial subsistence under-the-table employment, on public assistance, or unemployable, Gray's prenatally-induced and post-natally acquired brain damage left him less able to climb the social ladder toward the American dream. He had a hard time surviving in the underclass, much less in the middle class or upper middle class.

Is it possible that we as a society are simply tolerant, indifferent, numb, or ignorant to the obvious cause of the psychopathology in front of us? Is it simply too hard to bear the shame and guilt that our libidinal lubricant of choice is contributing to social problems that plague the already underserved, disenfranchised and marginalized among us? Today, while reading the *Journal of the American Medical Association*, I was encouraged by a young colleague, Brown University medical student, Katherine C. Brooks, who ventured into this realm of questioning, opening a dialogue about unspoken truths in plain sight that we deny or push into our subliminal collective unconscious. In her words,

> *In my medical training thus far, Trayvon Martin lost his life, Michael Brown was left to die in the streets of Ferguson, Missouri, and Eric Garner was choked by officers as he repeated eleven times that he could not breathe. But these events were rarely mentioned in the lecture hall, my small group sessions, or morning rounds. Was I supposed to ignore*

their implications for the lives of my patients and form my role as their caregiver?[56]

Ambivalence, head-shaking, and denial will surely arise from uncovering the veil of truth. Insights and optimism from my work with children, adolescents and adults who have ND-PAE provide hope that these young women and men can rise above the caste of the underclass. Similar to autistic disorders, early interventions and developmental disabilities services can improve ND-PAE as individuals mature through adolescence and transition to adulthood. Innovative approaches to identification, treatment and prevention of this condition will help transform lives, communities, and society as we know it.

There are a number of books written by individuals with effects of prenatal alcohol exposure, their parents, and professionals working with them that provide hope, optimism and strategies for successful outcomes. Some are listed in the Resources section at the end of this book. I have provided an evaluation and treatment planning paradigm in Appendix A that can help guide clinicians and parents in helping simplify the complex range of neurodevelopmental issues. It is clear from over four decades of research with these individuals that early diagnosis (prior to age 6) can be protective; however, many evaluators have told families that the diagnosis is irrelevant to treatment and would not benefit the individual's prognosis.

In time, true change will emerge by recognizing the human rights tragedy of productive lives lost due to victimization, homelessness, prostitution, sex trafficking, and substance use disorders of individuals with ND-PAE revealed in the next chapter entitled, *Predator ... or Prey?*

---

[56] Brooks KC. A Silent Curriculum. From A Piece of My Mind published in the *Journal of the American Medical Association*, May 19, 2015; Vol 313, No. 19: 1909-10.

Chapter 6

Predator...Or Prey?

Once John got in trouble for inappropriate touching while on a school playground. He was about 15 at the time, but emotionally at the level of a 5-year-old. ... The repercussions from this incident had a lasting impression on John, and invoked such fear into my heart for John's future, that I resolved to do everything in my power to keep John out of prison. So far I have succeeded. But only with close monitoring by either myself, his brother, his job coach, or his mentor. I can only hope that this success will continue after John enters a community home placement, which looms in the not-too-distant future as I prepare to apply for residential services for John. - Excerpt from online article by Teresa Kellerman, FAS and Inappropriate Sexual Behavior, Come Over to FAS, 2002.

"Us versus them" echoed off the shoes of the 6'3" muscular Goliath escorting me through the maze of bare linoleum, concrete walls and steel doors of the federal penitentiary. Shaking off a chill as though entering the isolation chambers of the epidemic's quarantine facilities, I passed into the brightly lit interview room. The guard nodded an unspoken greeting toward the detainee, beckoning me to a desk across from the short, dwarf-like man dressed in a bright orange jump suit. Donald Mendelson's facial features were out of a textbook on Fetal Alcohol Syndrome. His tiny, slanted almond shaped eyes were just a bit too widely spaced and lackluster even for my own comfort. Giving the impression of Batman's Joker, an absent Cupid's bow and barely visible upper lip stretched his wide grin into a plastic smile across his taught face. I shuddered softly, at once taken over by both anger and sadness at the tragedy of yet another life affected by the silent epidemic.

Moments like this give me pause for reflection. Years earlier, I had first learned to identify those same features as a young public health entrepreneur while working with a Native American pediatrician in rural NC. He taught me why Fetal Alcohol Syndrome ("dysmorphic" ND-PAE) alluded medical understanding for decades following prohibition's repeal. Most doctors simply labelled the strikingly odd appearance as "funny looking kid," or FLK for short. Pediatrician Dr. David Smith in at the University of Washington at Seattle developed the field of dysmorphology as a branch of medicine to study children with subtle abnormal facial features. Eventually he and colleague Kenneth Lyons Jones distinguished children with Fetal Alcohol Syndrome (FAS) as having a distinct set of features separate from other dysmorphic syndromes. Over the years leading up to their description of FAS, doctors would recognize similar features in one or both of the parents and incorrectly assume that the condition was genetic, not understanding the multigenerational transmission from mother to child by a neurotoxin she'd consumed. Noting the child's small, wide set eyes, button nose, flattened mid-face, and thin upper lip, the doctor would look at the parents, thinking – "They're pretty funny looking, too; so, it must be something genetic." Though "FLK's" tended to have minimal brain dysfunction, seizure disorders, autistic traits, and other forms of brain damage – no one thought to examine a culprit seemingly so benign and ubiquitous as prenatal alcohol exposure. My visceral response to the term FLK all those years ago compelled me to leave public health and pursue a career in medicine to provide more credibility to my message.[57]

Nearly 20 years later, I sat across from Donald Mendelson, a 30-something whose odd facial features, intellectual issues, and small stature caught the attention of his attorney with the Office of the Federal Public Defender. She learned from his father that he had been diagnosed twice with FAS by doctors at two separate teaching hospitals, first before puberty and later during his late 20's. Wondering how the diagnosis

---

[57] Rich SD, Fetal Alcohol Syndrome: A Preventable Tragedy. *Psychiatric News*; May 2005, Vol. 40, No. 9

might affect his culpability, the astute public defender called me in for a consultation. This was my first and only forensic case of an individual who had already been diagnosed with FAS prior to my involvement in the case. It was also the first time I had consulted involving sexual charges.

As I walked toward Donald, he halfway stood out of respect and deference, though in a contorted posture from being shackled and handcuffed to an adjacent pole. Not much taller than me, he appeared much younger than his chronological age and overly enthusiastic. Greeting him with a smile, I introduced myself before settling into a plastic chair at the table across from him. I explained the purpose of my assessment by request of his counsel and asked if they had discussed my visit. He said, "Yes, ma'am." I elaborate that my neuropsychiatric evaluation was for forensic purposes, meaning anything we discussed could be in a report to the court. Placing my instruments on the table in front of me, I made mental inventory to ensure they returned to the bag when I left – blood pressure cuff, ruler, stethoscope, tape measure, reflex hammer, and pen light. Nodding to my pen and legal pad, I asked if it would be okay to take notes and make some measurements of his height and other physical features. He again said, "Yes, ma'am," shifting in his seat to adjust his restrained arm, still tethered to the adjacent pole.

De je vu hit me sitting across from another person who on paper had significant learning disabilities that could have easily been identified in school age when community-based programs might have protected him from this outcome. If he had the face of Down's syndrome and not Fetal Alcohol Syndrome, more resources would have been rallied at his disposal from birth. Communities understand the face of Down's, never leaving the individuals to fend for themselves on the street. While individuals with ND-PAE may not have as profound deficits in measures of intellectual quotient (IQ) as those with some genetic disorders, their social and emotional intelligence and adaptive functioning often place them at the bottom tier of society. In my view, the fact that society chooses to lock them up and throw away the key because of their neurodevelopmental disabilities is no less than a human rights issue.

The most significant impact of brain damage associated with prenatal alcohol exposure is likely in a person's adaptive functions. Although their intellect may be average or just below average, they are often unable to access their intelligence to carry out activities of daily living, have age-appropriate social interactions, and apply academic knowledge to real life situations. A person's adaptive functions include abilities such as application of academic skills in daily life (balancing a checkbook, telling time, counting money, budgeting, sticking to a schedule), interacting and communicating in a socially competent way with typically developing peers (conversational style and content, receptive and expressive language, nonverbal cues, social pragmatics and perception, as well as understanding social nuances such as manipulative behavior by peers, and one's level of naivety, gullibility, and maturity. The third and final measure of adaptive functions is practical – including activities of daily living such as being able to care for one's hygiene, prepare nutritious meals, complete household chores, follow directions, maintain a household, and parent children. Donald's neuropsychological testing had rated him at the mild intellectual disability level in terms of adaptive functions. Although he had held low wage jobs not requiring thinking skills (such as a bicycle courier, a scripted telemarketer, and a construction worker), he was unable to manage his money, balance a checkbook, or maintain relationships with higher functioning peers.

Because his father, Leonard Mendelson, was available and willing to speak with me, I had interviewed him for collateral history before the diagnostic assessment with Donald. I learned that his mother, Josie Branch, became pregnant unexpectedly while she and Leonard were dating. Being "the party type," the couple regularly used alcohol before and after they realized she was pregnant. Josie's sister corroborated that Josie frequently drank to the point of intoxication throughout her pregnancy with Donald. Though she carried the baby full term and had a seemingly uncomplicated delivery, Donald was born slightly premature and below average birth weight. Leonard recalled that Donald "thrashed around in the incubator" after birth and was not an easy baby to comfort. He was difficult to soothe, irritable, had feeding difficulties, and did not sleep well.

Leonard and Josie argued and fought during much of Donald's early years. As a mother, Josie was ill equipped to care for her young son, staying drunk much of the day while Leonard was working. Often, Leonard returned home from work finding Josie drunk and passed out on the sofa with Donald wandering around the apartment, unkempt and uncared for. When Donald was a toddler, Leonard recalled an incident in which Josie called him at work concerned that their son had disappeared earlier that day. She had been drinking and not aware he had opened the door and left the apartment. Driving home in a panic, Leonard found Donald wandering in the neighborhood wearing only a diaper and tee shirt in the middle of the winter. The couple eventually split up, leaving Donald with years of significant abuse and neglect. Eventually, Child Protective Services removed him from the home and placed him with his father in early elementary school. At that point, he went to live full time with his father and has not seen or had contact with his mother since.

Mr. Mendelson described Donald as shy and anxious, though a warm, loving, and empathetic person. From an early age, he struggled with delays in talking, walking, and other developmental milestones. Donald saw early intervention specialists through the county's Infants and Toddlers Program and was treated by his pediatrician with Attention Deficit Hyperactivity Disorder (ADHD). No one ever inquired about prenatal alcohol exposure or related problems. Parenting was challenging due to his disruptive behaviors. He tended to run around, climb on furniture, and behave like a bull in a China shop, even in public places. Like many other children with ND-PAE, he was dismissed from ("kicked out" of) more than one nursery school due to hyperactivity and impulse control problems. In elementary school, he had difficulty following directions due to receptive language problems, distractibility, hyperactivity ("Ever Ready bunny" behavior), and working memory issues. Because of learning disabilities, it was also hard for Donald to absorb information, read, or learn math. Unfortunately, none of his teachers were aware that he had prenatal alcohol exposure (although he had all the features of Fetal Alcohol Syndrome). He was short in stature, had characteristic facial features, and exhibited the neurodevelopmental problems of a child with FAS.

Teachers described him as having social communication issues – not understanding non-verbal cues, having trouble with peers, and problems with language pragmatics – understanding the back and forth of conversation. He had few friends and preferred spending time with either younger children or adults due to anxiety and social skills problems. His father took him to psychiatrists and a number of therapists due to his immature behaviors, hyperactivity, and school difficulties. Due to his small size and immature behaviors, he was bullied relentlessly by peers and felt socially isolated throughout his school years. School psychological tests estimated he was years behind his peers in social and emotional development. Yet no one in his life – teachers, principals, caregivers, health care providers, or his parents connected his symptoms to a form of brain damage due to prenatal alcohol exposure.

As an adult, Donald gained insight that other people took advantage of him and he was gullible. He admitted trusting people and being coerced into unlawful activities, but then felt betrayed when he would be the one to get in trouble. He remembers his father being concerned about the type of people he was hanging around, and that he often took the blame for others. He was extremely loyal to his friends and would never betray their trust by "ratting them out." Often, he was the only one left at the scene of a prank, a party, or a dare and tended to get caught in the kind of foolishness that other kids masterminded and he implemented. From an early age, these vulnerable children are seen as the scape goat and grow into adolescents and adults with ND-PAE are frequently targeted to be mules for drug dealers, "spotters" for robberies, and co-defendants in crimes.

Being extremely small for his age since he was a toddler, concerns about his growth led his father to seek help at a university medical center for an evaluation before puberty. At the time, Donald was in fourth grade, below four feet tall and weighed less than forty pounds – clinically about the average weight for a four-and-a-half-year-old boy and height as well as bone age of a 6-year-old boy. An endocrinologist diagnosed Donald with failure to thrive, which meant that he was not achieving his full growth potential. A complete pituitary evaluation showed that he was producing adequate growth hormone but supernormal growth hormone response. Subsequent testing ruled out genetic forms of growth hormone deficiency, which led to a second opinion from a second medical center department of endocrinology and metabolism. In the end, Donald received growth hormone before and during puberty as a treatment for his growth failure.

The hospital endocrinologist also referred him for a multidisciplinary, comprehensive assessment in the department of pediatric genetics where he was diagnosed with Fetal Alcohol Syndrome (FAS). Even as an adult, Donald had characteristic facial features of small, wide set eyes; short, upturned nose; a thin upper lip, and relatively smooth philtrum (groove between the nose and upper lip). He also had the other features of FAS, including growth retardation and neurodevelopmental deficits. Unfortunately, the genetic pediatrician mistakenly told Leonard Mendelson there was no point to tell Donald he had FAS and that therapy would not be warranted or helpful. Other than receiving the growth hormone therapy (daily injections for a number of years), he did not receive any other therapy based on the diagnoses of growth deficiency or FAS.

By high school, Donald had been diagnosed with a number of other neurodevelopmental and neuropsychiatric issues, including learning disabilities, ADHD, anxiety, and depression. His father felt Donald's chronic depression was related to his difficulty developing lasting friendships; early neglect, abuse, and lack of affection from his biological mother; and academic struggles leading to the need for special education services. During middle school, he witnessed boys from the neighborhood having sex with a girl while they were all drinking alcohol at a friend's house. He denied participating but said that the incident was disturbing to him. After being expelled from several public high schools, he was eventually enrolled in a therapeutic treatment center then was transferred to a day program. Eventually, he was pushed through to graduation although he could barely read or write. So often in my practice I have seen young adults who have graduated from a therapeutic high school for individuals with severe learning issues but unable to write a coherent paragraph even with multiple misspellings and run-on sentences. Like Donald, these issues put the young adults at risk of failing at jobs because employers assume that they are high school graduates and capable of a higher level of functioning than they are realistically able to do.

Since high school, he has worked in a variety of low skill, menial jobs requiring limited conceptual skills. A strong work ethic and fairly good telephone skills helped him be proficient in a political campaign call center. He enjoyed working but never felt he fit in with his co-workers due to always feeling much younger than his chronological age. He was unable to keep up the expenses of his apartment or handle managing his own finances, which is often the case with young adults living with ND-PAE. Often his father lent him money to pay bills that were overdue and to overcome financial difficulty. He rarely socialized with people at work and had few friends other than the few women he dated. He has rarely drunk alcohol more than a few social drinks and has not used substances beyond experimentation in high school with marijuana.

Donald had a couple of long term relationships in which the women took over custodial care of running the apartment, managing their money, cooking and cleaning, and keeping the place tidy. One of his girlfriends complained that he seemed distant and aloof to her feelings, and he later acknowledged that he has difficulty feeling emotions. While living with one woman, he made "stupid sexual comments" to other women, joked around inappropriately, and met several women on-line, some older than him and some younger. He "cheated" numerous times with other women, even though he says that he loved the one he was with. He met his last girlfriend while dating her best friend. Both his girlfriends were aware he has disabilities and had emotional issues of their own (i.e., depression, family issues, and difficulties with school.

Later, after Donald began to have trouble with the justice system, his father recalled the original diagnosis and believed that Donald should be re-evaluated to see if prenatal alcohol exposure might have affected his behaviors. A geneticist at a multi-disciplinary FASD program assessed him and confirmed the diagnosis of FAS. A chromosome microarray ruled out genetic disorders as a cause of his growth retardation, abnormal facial features, and neurodevelopmental disabilities. The staff recommended a therapist to help Donald learn adaptive functions and provide parent guidance to his father to help apply for community services. They also encouraged Donald to wear a medic alert bracelet indicating he has FAS and is unable to waive his Miranda rights.

As with other forensic cases, I typically do not look into the client's legal issues until after the interview. I knew from discussions with his public defender that he was in jail for sex with a minor. Because of his immature social and emotional behaviors as well as the stature and adolescent appearance, Donald has always interacted with peers much younger than his chronological age. Donald's gullibility and appearance played a role in his being perceived as in his teens though he was in his early 30's. He had moved back in with his father after being released with an ankle bracelet after a similar offense a few years earlier. He was lonely and once again started using online dating sites where he met a high school student posing as an 18-year-old. Although the girl was not yet 16 at the time, her well-developed physique and misleading messages convinced him that she was of legal age. As a result of his loneliness, searching in the wrong places for companionship, poor insight and judgement, Donald Mendelson was convicted of federal criminal sexual charges against two minors. A link between sex offenses and ND-PAE is becoming more recognized as therapists and professionals working with individuals convicted of those crimes become educated about the effects of prenatal alcohol exposure on the development of the brain and sexual organs.

ND-PAE and Sexually Dimorphic Features

In her book, *Silent Spring*, Rachel Carson delivered a powerful message about the devastating effects of environmental chemicals on birds, plants, and other wildlife – cautioning policy makers and chemical manufacturers about the potential harm to our natural world from spraying for mosquitoes, other pests, and weeds. Some of these chemicals, known as "pseudo-estrogens"[58] or environmental

---

[58] Pseudo-estrogens are substances that cause effects like estrogen in the body. Alcohol has estrogenic effects, effeminizing the male and leading to androgynous effects in the female

estrogens,[59] have been found in our drinking water from factory waste and farm runoff containing pesticides and herbicides. These chemicals act by mimicking the female hormone, estrogen, interfering with gender differentiation of the brain and development of secondary sexual characteristics.

Similarly, alcohol acts in humans and animals as an "endocrine disrupter," leading to widespread effects on the developing fetus, child, adolescent, or adult exposed to the chemical solvent. In healthy adult women, moderate to heavy alcohol use can effect regularity of menstrual cycles, decrease fertility by affecting ovulation (release of the egg), and increase rates of reproductive cancers. In adult males, frequent alcohol use can decrease the male hormone, testosterone, cause gynecomastia (development of breasts), and lead to sexual dysfunction.[60]

Invisible and unseen, the stealth chemical also interferes with hormone regulation during prenatal development in both males and females. Infant boys exposed to alcohol during pregnancy can have feminization of the male genitalia, body, and brain. Some are born with their penile opening in the wrong place (hypospadias), have small gonads (sex organs), or have effeminized brains. It is no wonder then that many young adults with ND-PAE end up with gender identity issues. A notable case in recent years is Bradley (now Chelsey) Manning – Wikileaks defendant who was diagnosed with Fetal Alcohol Syndrome by one of the psychiatrists on the team evaluating her/him.[61] Transgender issues eventually led to his having a sex change operation.

As an endocrine disrupter, alcohol effects the hormone systems in the body during development and throughout a person's life. The reproductive hormones work on babies from before birth all the way through childhood, adolescence and adulthood – regulating development

---

(Gavaler JS. Alcoholic Beverages as a Source of Estrogens. *Alcohol Health and Research World;* Vol. 22, No. 3, 1998, p. 220-227).

[59] Environmental estrogens are chemicals that mimic estrogen, such as certain pesticides, herbicides, plant derivatives, and oral contraceptives.

[60] Emanuele N and Emanuele MA. The Endocrine System: Alcohol Alters Critical Hormonal Balance. *Alcohol Health and Research World;* Vol. 21, No. 1, 1997.

[61] McKelvey T. Bradley Manning's disrupted family life. *BBC News Magazine;* 22 August 2013. http://www.bbc.com/news/magazine-23780581

and maturation of the brain, body, and sexual functions. Babies exposed to alcohol prenatally are at risk of birth defects of the sex organs and urinary tract,[62],[63] such as differences in the placement of the penile opening (a condition called "hypospadias"),[64] feminization of the brain and gonads (causing smaller testicles and penis), and delayed puberty onset.[65] Growth problems have been seen in adolescents exposed to as little as 1 drink per day during pregnancy.[66] Like Donald Mendelson, they can also have a higher pitched voice, immature social behaviors, and a propensity for sexual indiscretions. Feeling more comfortable with and relating to younger peers, individuals with ND-PAE are often developmentally more similar to peers several years younger.

In my own clinical practice, I have worked with a number of patients with ND-PAE who dress and act androgynous – males appearing effeminate and females appearing masculinized. Concerned parents from other states have contacted me to ask what I know about the link between gender identity and/or dysphoria and prenatal alcohol exposure. Animal studies have shown a link between prenatal alcohol exposure and masculinization of females as well as feminization of males.[67] Studies of differences in sexual preferences between prenatally exposed and non-exposed male animals[68] suggest that these changes may be related to a decrease in size of the hypothalamus, a brain structure that is normally larger in males than females. Prenatal alcohol exposure has been shown to interfere with sexual differentiation by altering the

---

[62] Abel EL: *Fetal Alcohol Syndrome and Fetal Alcohol Effects.* New York, Plenum Press, 1984.

[63] Sokol RJ, Miller SI, Reed G: Alcohol abuse during pregnancy: An epidemiologic study. *Alcoholism*; 1980;4:135-145.

[64] Mills JL and Graubard BI. Is Moderate Drinking During Pregnancy Associated With an Increased Risk for Malformations? *Pediatrics;* September, 1987; 80: 309-314.

[65] Carter RC, Jacobson JL, Dodge NC, Granger DA, Jacobson SW. Effects of prenatal alcohol exposure on testosterone and pubertal development. *Alcohol Clin Exp Res.* 2014 Jun;38(6):1671-9.

[66] Day NL, Leech SL, Richardson GA, Cornelius MD, Robles N and Larkby C. Prenatal Alcohol Exposure Predicts Continued Deficits in Offspring Size at 14 Years of Age. *Alcoholism: Clinical and Experimental Research.* Volume 26, Issue 10, pages 1584–1591, October 2002.

[67] McGivern RF, Clancy AN, Hill MA, Noble EP. Prenatal alcohol exposure alters adult expression of sexually dimorphic behavior in the rat. *Science.* 1984 May 25;224(4651):896-8.

[68] Dahlgren IL1, Matuszczyk JV, Hård E. Sexual orientation in male rats prenatally exposed to ethanol. *Neurotoxicology and Teratology.* 1991 May-Jun;13(3):267-9.

male hormones (androgens)[69] (i.e., lowering testosterone production)[70] similarly to effects in adults.

ND-PAE and Premature Sexual Arousal

I recently have begun seeing two adopted African American boys, both with heavy prenatal alcohol exposure, from the inner city of Baltimore – now living in suburban Maryland. One is 20 years old and the other just over 12. Years ago, the older one had been touched inappropriately in their home by two brothers who were there for foster care. He began having sexual play with his middle brother after the two foster children left, then the middle brother (not my patient) started sexually stimulating the younger one.

When the brain is stimulated from an early age toward sexuality, neurons are activated and become hyper-aroused in such a way to then cause the person to seek out sexual stimulation. Over time, the types of stimulation are not enough and the person begins to experiment with other methods to arouse themselves – not just from soft porn to hard porn magazines to Internet porn to phone sex and beyond. Many children with neurodevelopmental limitations have strong libidinal drives that lack the inhibitions of higher level cortical processes. From early ages, parents are appalled that children with ND-PAE touch themselves, masturbate frequently, and explore sexuality with peers. When sexually aroused at an early age, they are more likely to develop sexual behaviors that appear deviant or immoral. Many are emotionally and socially much less mature than their chronological age; so, their sexual indiscretions with developmentally similar peers (still considered minors) often get them into trouble with the law, as was the case with Donald Mendelson and countless others like him.

Remember the hedonistic nurseries of Aldos Huxley's *Brave New World*? Epsilons and Deltas (exposed prenatally to alcohol) were

---

[69] Barron S, Bliss-Tiernan S, and Riley EP. Effects of Prenatal Alcohol Exposure on the Sexually Dimorphic Nucleus of the Preoptic Area of the Hypothalamus in Male and Female Rats. *Alcoholism: Clinical and Experimental Research*. Volume 12, Issue 1, pages 59–64, February 1988.

[70] Hård E, Dahlgren IL, Engel J, Larsson K, Liljequist S, Lindh AS, Musi B. Development of sexual behavior in prenatally ethanol-exposed rats. *Drug and Alcohol Dependence*. 1984 Sep;14(1):51-61.

encouraged to have sexual play at an early age, running around the playground playing hide and seek in the bushes, touching each other's private areas and chastised when they complained. A case in point is 16-year-old Stephanie, who has been at a residential treatment institution in Maryland for the past 18+ months. Her biological Caucasian mother, Stella, was 16 years old and in foster care herself when Stephanie was born. She and Stephanie's father, a Hispanic man a few years older, had a second child, Bella, now 13 years old and also my patient. Stella was documented by Child Protective Services (CPS) as having abused alcohol throughout both pregnancies. When Bella was an infant and Stephanie was 3-½, CPS removed them from the home because of profound neglect and abuse by Stella's new boyfriend, who watched pornography videos all day in their government subsidized apartment. According to reports, the couple also frequently had sex in the same room where the children played.

For the year they were placed in foster care as well as throughout childhood after being adopted by Evelyn and Harvey Jenkins, Stephanie displayed hyper-sexuality – frequently masturbating while sitting by herself or with others watching television. At school, she would prop herself up on the corner of a desk or chair and stimulate to the point of orgasm so frequently that she was being teased and mimicked by peers. The Jenkins attempted to dissuade the behavior by removing anything with a corner from her bedroom (desk, dresser, night stand, and table). Even at the age of sixteen, she uses self-stimulation in times of stress, excitement or boredom to self-soothe and comfort. Early exposure to sexuality can have profound and lasting effects on the psyche – particularly for individuals whose brains are libidinally-driven and limbically-wired.

Individuals with prenatal alcohol exposure can have heightened libidinal drives, particularly when witnessing or experiencing early sexual stimulation or sexual trauma. Often, judges and prosecutors will inaccurately assess the person to be a menace to society and incapable of rehabilitation, like the judge in the case of Donald Mendelson. More of these young people need parents who understand their limitations and can advocate for and protect them from wrong-doing. Parental supervision, or supervision by a guardian, is often the only way to help ensure that they do not unwittingly incriminate themselves by behaviors seen as antisocial, predatory, or sexually provocative.

In her landmark longitudinal study of 451 patients of all ages, Dr. Anne Streissguth of the University of Washington at Seattle found children and adults with prenatal alcohol exposure have a history of the following sexual behaviors (starting with the most prevalent): sexual advances, sexual touching, promiscuity, exposure, compulsions, voyeurism, masturbation in public, incest, and obscene phone calls.[71] Teresa Kellerman cites the Streissguth article in her online newsletter as she describes her son, John's inadvertent sexual behaviors:

> *Those with an IQ under 70 tend to get into trouble for inappropriate sexual touching and for masturbating in public. We all know that means they touch or rub themselves through their clothes without even thinking, in spite of our repeated reminders that this is not okay behavior in public. And of course, the sexual touching is not always intentional or understood as being wrong at the moment.*
>
> *According to Streissguth, the females in her study who had sexual behavior problems were more likely to have been victims of sexual abuse. The males in her study with sexual behavior problems were more likely to get into trouble with the law. Among both males and females, those who had been victims of violence were four times as likely to exhibit inappropriate sexual behavior as those who had not experienced violence. John has never experienced or witnessed violence, nor had he ever been neglected or abused as a child, emotionally, sexually, or any other way. Still, at age 10, if he were in a room with other children who had been abused, neglected, or bounced from home to home, you would not be able to pick him out, as his behavior looked just as if he had been sexually abused and had never been taught proper manners. I'm afraid at age 24, there are times he still gives this impression.*

---

[71] Streissguth, AP, Bookstein FL, Barr HM, Sampson PD, O'Malley K, and Kogan Young J. Risk Factors for Adverse Life Outcomes in Fetal Alcohol Syndrome and Fetal Alcohol Effects. *Developmental and Behavioral Pediatrics.* Vol. 25, No. 4, August 2004.

Once John got in trouble for inappropriate touching while on a school playground. He was about 15 at the time, but emotionally at the level of a 5-year-old. It's a long story, that I will tell sometime soon. The repercussions from this incident had a lasting impression on John, and invoked such fear into my heart for John's future, that I resolved to do everything in my power to keep John out of prison. So far I have succeeded. But only with close monitoring by either myself, his brother, his job coach, or his mentor. I can only hope that this success will continue after John enters a community home placement, which looms in the not-too-distant future as I prepare to apply for residential services for John.

In the meantime, I will continue to monitor John closely, especially in social situations, trying not to feel guilty about depriving him of a "normal life of independence" and trying not to buckle under the pressure professionals impose in the name of self-determination for me to just "let go." If I gave John the freedom that some disability experts think I should, it would only be a matter of time before he became another statistic in the criminal justice system. We could certainly let our child-adults fly from the nest, if it weren't for their Broken Beaks and Wobbly Wings.[72]

After reaching out to Teresa for her permission to use the excerpt from her website, she sent me the following postscript:

My son John is 38 now, living in an apartment and holding down a job. He has 24/7 supervision and is closely monitored so he will hopefully not get into trouble. He has not been in trouble with the law, but he has had a few close calls. For instance, at his apartment complex, while with his mentor in the swimming pool area, he was staring at a 9-year-old girl (probably wanting to be her friend), and she went home and told her dad, and her dad called the police. I arrived before the police did, and I didn't allow the police officer talk to John, but instead convinced him that John's being friendly was innocent and intended nothing harmful to the girl.

---

[72] Kellerman T. FAS and Inappropriate Sexual Behavior; online article. *Come Over to FAS*; © 2002.

> *This kind of situation is very scary for John. He does not want to end up behind bars, but he cannot control his impulse to flirt and stare and be friendly to women, which can easily be misinterpreted as sexual harassment. John is not interested in women sexually, but just wants them to like him and accept him. He is a sweet and loving person. He just wants what everyone else wants, to be loved and accepted and to be part of the group. He wants to be respected and he works very hard at being respectful. Having a Fetal Alcohol Spectrum Disorder means he continues to struggle with impulsivity and social immaturity. It's a challenge he will face daily the rest of his life.*

Prey to Victimization

One of the human rights tragedies of this epidemic is the effect of prenatal alcohol exposure on vulnerability and gullibility to both prey and predatory behavior. LaDonna Simpson was 15 years old when a trusted adult introduced her to a 55-year-old man living in a nearby city across state lines. Having been adopted from Russia at age 5 by a couple in their 40s living near the Greater Washington D.C. metropolitan area, she was in a special needs private school receiving intensive academic services for learning disabilities. Over the next 4 years, the man seduced her with expensive gifts, leaving track phones on her parents' back porch so that he could keep in touch with her. He talked her through the process of bypassing security passwords on her parents' computers so that they wouldn't find out about the affair. When she graduated high school at age 19 and failed her first semester of community college, he convinced her to "elope" with him– picking her up from her house at 4 in the morning and driving to Florida where they were married by a justice of the peace. Returning to suburban D.C., he kept her in his townhouse for the next month, having her do online pornography – pretending to be his "naughty daughter" who needed spanking. At 4'7" and 90 pounds, LaDonna succumbed to his wishes, too scared and submissive to reach out for help.

LaDonna's distraught parents eventually found her through a private detective who knocked on the door one day while her "husband" was at work. Disheveled and frightened, yet sobbing with joy for the "rescue,"

LaDonna quickly gathered her things and fled with the detective, relieved to be out of the situation. Over the intervening months, she missed having the "nice things" – expensive clothing and jewelry he gave her, and often thought about returning to live with him. She was psychologically scarred by the experience, feeling shame and guilt over the actions she was forced to do. Despite the ambivalence, with her parents' help, LaDonna was able to get a divorce from the man, who relented only so that she and her parents wouldn't press charges against him.

Like other individuals with ND-PAE, LaDonna had always struggled academically despite inordinate supports in school. She failed community college a second time after returning to live with her parents and began to steal money and items from their home to pawn, encouraged by "friends" she met who were also failing school. Eventually, she and her parents decided to have her go live on a ranch with her younger sister, who was also adopted from Russia at age 2 ½ and has ND-PAE. There they have affordable supportive housing in a small group home on a 90-acre ranch with staff who understand their disabilities and provide assistance with activities of daily living, transportation to appointments and access to recreational activities in the community. LaDonna enjoys theater and art, so she has been able to perform in plays at the local community theater and spends her free time drawing and painting. She and her sister work with the other residents to take care of the animals – raising calves donated from the local dairy to be sold as beef in order to support their living expenses. This type of housing in a cooperative vocational program is what individuals with ND-PAE need to prevent them from being duped, molested, and mistreated. The safe, structured environment provides the external supports that LaDonna and her sister need in order to live happy, productive lives independent of their parents.[73]

---

[73] See YouTube video – Elizabeth H, FASD: https://www.google.com/url?sa=t&rct=j&q&esrc=s&source=web&cd=1&cad=rja&uact=8&ved=0CB4QtwIwAA&url=http%3A%2F%2Fwww.youtube.com%2Fwatch%3Fv%3Db02BK13zQBY&ei=V_cEVYfoKMmggwS1rYTgBw&usg=AFQjCNEtV9qP3XYIsrXi_GzJH7swIvr0Lg&sig2=OEhqM0dcjavXsutriUhrCg&bvm=bv.88198703,d.eXY.

So often we read in the headlines stories of young women, teens and children being kept in basements for years as the "sex slaves," forced laborers, or housekeepers of one or more people living in the home. We ask ourselves, "Who would allow someone to do that to them? What would cause a person to remain enslaved by another human being? Why do they stay under such depraved conditions?" The more I have learned about the struggles of individuals with undiagnosed ND-PAE, the clearer the answers are to such questions. In my view, the highest risk individuals are those with ND-PAE – gullible, easily influenced by peers often running away from abusive families, lacking social judgment or a vocational skill set to gain and maintain a job with wages to support themselves. Their biological parents may be ill-equipped to intervene when other adults take advantage of or kidnap their children or adolescents. Thus, they filter down through the social pyramid to the murky waters of the underclass.

One of the agencies founded to make a difference for these young women is Rights4Girls, based in Washington, DC (www.Rights4Girls.org). This agency estimates that 100,000 American children are at risk of child sex trafficking each year and that 12- to 14-year-old girls are the highest risk group of commercial sexual exploitation. In order to shift this human rights tragedy toward social benevolence and protection of the innocent, we must strive to create intentionally supportive communities – where people with hidden disabilities can thrive, not just survive. They deserve to have respectable lives with meaning, purpose, and fulfillment – protected from others who would take advantage of them.

In the next chapter on *Historical Perspectives,* the science behind responsibility to our future generations echoes from seven generations of lost, then reclaimed, knowledge.

Chapter 7

HISTORICAL PERSPECTIVES: ILLUMINATING THE PAST TO ENLIGHTEN THE FUTURE

Those nations of men that have used the strongest alcoholic beverages through many generations have now, through a standpoint of performance and modern accomplishments, have outstripped the other nations with less alcoholism in their history. This may be due to some such selective effects as those recorded in the foregoing pages."[74]

— *Charles R. Stockard, April 1924.*

As a science enthusiast all through my childhood and adolescence, I was drawn to the study of biology rather than history, political science, or the social sciences. Nonetheless, lessons from history teach us that we are discovering and rediscovering information about alcohol's ill-effects on human development and lost human potential. By learning from the past, we can weigh societal costs and put in place systems to prevent prenatal alcohol exposure from dulling the minds of the unborn. While a comprehensive historical literature review is beyond the scope of this book, insight into the history of the ebb and flow of knowledge over the past few centuries is relevant to understand the silenced science behind the epidemic.

---

[74] Stockard, CR. Alcohol a Factor in Eliminating Racial Degeneracy; *The American Journal of the Medical Sciences*; April 1924.

While most epidemics rear their ugly heads with the gore of leprosy, the scourge of Black Plague, the devastation of tuberculosis, or the hidden terror of HIV, ND-PAE is neither an obvious nor a communicable illness. It is a silent menace in society's underbelly, left unchecked and unrecognized to the masses for as long as mankind has imbibed alcohol – only to flourish and reach epidemic proportion at times when social pressures and market forces allow alcohol to flow freely without regard for health.

Some scholars cite Biblical warnings about alcohol use around the time of conception; although European societies – Vikings, Celts, English, Germans, French, Italians and Irish – exposed countless generations to their heavy alcohol consumption. In her book, *Message in a Bottle: the Making of Fetal Alcohol Syndrome,* Janet Golden provides a far deeper appreciation of the socioeconomic and historical context for understanding of the teratogenic potential of alcohol.[75] Her well-researched book provides details about the social and political forces competing to sway majority public opinion as American scholars realized our beloved intoxicant was capable of causing harm to the unborn.

During the 18th century, alcohol abuse became rampant in England with British soldiers bringing the cheap beverage, gin, from the Netherlands. A combination of forces contributed to the ensuing epidemic of gin consumption – laws discouraging brandy imports from France and a breaking up of the London Guild of Distillers in 1690. The cost of food dropped in Great Britain around that time, leaving consumers with funds to purchase alcohol. Overconsumption of alcohol led to a dramatic rise in alcohol abuse and a response by moralists and legislators tantamount to the War on Drugs.[76] While the educated, affluent drank beer – not seen as causing pandemonium and mayhem, the less fortunate drank the much cheaper gin.[77] Author Daniel Defoe commented:

---

[75] Golden J. *Message in a Bottle: The Making of Fetal Alcohol Syndrome.* Harvard University Press. 2005.
[76] https://en.wikipedia.org/wiki/Gin_Craze
[77] Ernest L. Abel. The Gin Epidemic: Much Ado about What? *Alcohol and Alcoholism*; Volume 36, Issue 5; pp. 401 – 405.

> ... the Distillers have found out a way to hit the palate of the Poor, by their new fashion'd compound Waters called Geneva, so that the common People seem not to value the French-brandy as usual, and even not to desire it.[78]

Advocacy efforts promoted by Henry Fielding in *An Enquiry into the Causes of the Late Increase of Robbers,* underscored the contribution of freely flowing gin to social unrest and crime. The complementary etching by William Hargoth (*Gin Lane*, Figure 9) depicts living zombies drinking themselves to death, coffin houses as common on the streets as Starbucks shops or churches today, and children left to fight with dogs for scraps of food and bones. In the foreground, an inebriated prostitute hired as a wet nurse is unaware that the baby she is suckling has fallen to its death in the stairwell, while she reaches for a dip of snuff. In his plea to Parliament to end the gin craze, Fielding cited concerns about children conceived under the influence of distilled spirits. Scholarly papers have detailed the historical background and socio-cultural perspectives of the policy makers of that era, debating more the class bias than the knowledge that alcohol use was causing harm to infants and children.[79,80] Eventually, policies were enacted by Parliament that deterred the production and sales of gin, ending its impact on the impoverished masses.[81,82,83]

---

[78] Defoe D. *The Complete English Tradesman*, Vol 2, Page 91, 1727.
[79] Ernest L. Abel. Gin Lane: did Hogarth know about fetal alcohol syndrome? *Alcohol and Alcoholism*; Volume 36, Issue 2; pp. 131 – 134,
[80] Elizabeth M. Armstrong and Ernest L. Abel. Fetal Alcohol Syndrome: The Origins of a Moral Panic. *Alcohol and Alcoholism*; Volume 35, Issue 3; pp. 276 – 282.
[81] Wikipedia citation to Kate Chisholm (2002-06-09). "A tonic for the nation". Telegraph. Retrieved 2010-08-30. In a review of *The Much-Lamented Death of Madam Geneva* by Patrick Dillon, August 8, 2013.
[82] Abraham John Valpy. *The Pamphleteer, Volume 29*.
[83] Wikipedia, http://en.wikipedia.org/wiki/Gin_Craze.

Figure 8. Gin Lane by William Hargoth, 1751.

What must become an infant who is conceived in Gin ... with the poisonous distillations of which it is nourished, both in the Womb and at the Breast?[84]

In the years following the Revolutionary War, during reconstruction and industrialization, the American lifestyle began to involve excessive drinking due to the social stressors of inflation, hard physical labor, and easy access. Not only were workers often paid in part with alcohol, fermented and distilled alcohol was often considered a healthier beverage due to contamination of water. In certain areas, it was a customary part of social gatherings from marriages to elections to military functions. During the late 1800s, public health authorities ranging from alcohol researchers to physicians began to publish reports in journals and present findings at American and international conferences linking parental alcohol use to "degeneracy" (i.e., intellectual, physical, and neurodevelopmental problems) in offspring. There is evidence that such public health knowledge and data about harmful effects of alcohol on

---

[84] *An enquiry into the causes of the late increase of robbers.* Henry Fielding. London: A. Millar; 1751.

infants and children, gathered in the late 19th and early 20th century, contributed to Prohibition policies in the United States. A variety of studies, published in scientific journals and presented at international conferences, highlighted alcohol as a cause of degeneracy in children, stillbirth, and early childhood mortality. In turn, such knowledge influenced the Temperance Movement and national Prohibition of alcohol from 1920 to 1933.

Researchers into infant and child mortality and morbidity were analyzing statistical data during the early 1900s as a way to understand the influence of alcohol use on human health. These scholars were attempting to link maternal alcohol use with poor infant and child health and mortality. One of these alcohol researchers, a prominent physician, W.C. Sullivan, M.D. published a comparative analysis of sober (non-drinking) mothers and those drinking alcohol in 1906. He found that more than half the babies died among drinkers whereas less than one-quarter died among sober mothers. Dr. Sullivan also looked at 444 children born to 120 "alcoholic mothers." He found that 33.7% died from the 80 "first born children" in the cohort; 50% of the second born died; 52.6% of the third born died; 65.7% of the fourth and fifth born; and 72% of the sixth to tenth born. He also found that 4.1% of surviving children had epilepsy (seizures) and others were "mentally defective." It is unclear from the analysis at what age his review of the birth and death records included and what a comparable cohort of nondrinkers' infant mortality would look like. He inferred from his findings that "deaths of babies increased as mothers became more alcoholized."

Deaths of Babies Increased as Mothers Became More Alcoholized

Studies of 444 Children of 120 Alcoholic Mothers

First Born 80 Children: 33.7% Died
Second Born 80 Children: 50% Died
Third Born 80 Children: 52.6% Died
Fourth and Fifth Born 111 Children: 65.7% Died
Sixth to Tenth Born 93 Children: 72% Died

Of the Living Children 4.1% were Epileptic
Others were Mentally Defective

Stoddard CF. *Handbook on Modern Facts About Alcohol*. The Scientific Temperance Society, Boston, 1914.

Another study reported at the *Laitinen XII International Congress on Alcoholism* in 1909 cited differences in child mortality in drinkers' families compared with non-drinkers' families. A total of 19,519 children and 5,735 families were included in the study. Parents who abstained were those who had consumed no alcoholic beverages (at least since marriage). "Moderate drinkers" were those who had consumed no more than one beer daily (considered light drinking by today's standards) and "immoderate drinkers" drank more than one beer daily (a wide range – may be considered moderate to heavy use today). The child death rate in the abstaining families was 13%, in the moderate drinker families was 23%, and in the immoderate drinking families was 32%.

> Laitinen XII
> International Congress on
> Alcoholism, 1909
>
> "Statistics of 19,519 children in 5,735 families"

Child Death Rate Higher in Drinkers' Families

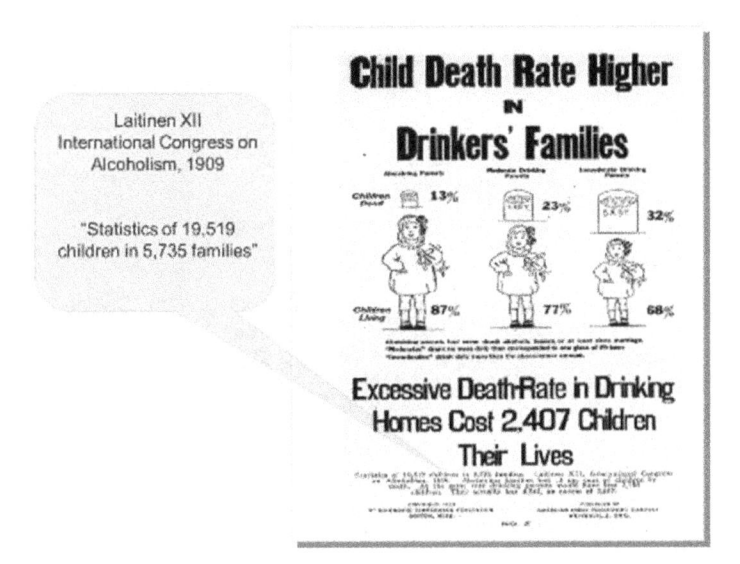

13% 23% 32%

87% 77% 68%

Excessive Death-Rate in Drinking Homes Cost 2,407 Children Their Lives

Stoddard CF. *Handbook on Modern Facts About Alcohol*. The Scientific Temperance Society, Boston, 1914.

A separate study published in an Austrian journal reported a much smaller cohort of children and families. Their study compared 120 sober families with 650 children and 18 beer drinking families with 125 children (all "strictly comparable and free of disease). In the village studied, beer was "practically the only drink used by the parents and not always immoderately." Sober parents lost 23% of their children, with 18.6% in the first year of life whereas beer drinkers lost 45% of their children, with 36% in the first year of life. While the study would not be scientifically or epidemiologically valid by today's standards, the researchers did point to beer drinking as a reason for infant and child mortality. Such research influenced the fervor of the Temperance Movement, which was driven largely by religious organizations. This is perhaps the key reason that the information about the effects of alcohol on children was "lost" to pre-prohibition antiquity – that the research was highly moralistic in tone.

> **WHY AMERICA WENT DRY**
> **Beer Doubled The Child Death-Rate**
> IN THE FIRST FIVE YEARS OF LIFE
>
> Austria, 1914
>
> Children of Sober Parents **23% DIED**
>
> Children of Beer Drinkers **45% DIED**
>
> Alcohol whether in Beer or in Whisky is an Enemy to Child Life.

Stoddard CF. *Handbook on Modern Facts About Alcohol.* The Scientific Temperance Society, Boston, 1914.

At that time in history, Darwinian scientists were pursuing the effects of genetics and environment on trait development. One of the earliest teratologists (scientists who study birth defects), studying the effects of prenatal alcohol on development in animals was Dr. Charles Stockard, whose research in guinea pigs and fish showed dramatic effects even several generations after the initial exposure – even if the first exposure was the father. If the parent guinea pigs were exposed 6 days per week to alcohol for several generations, he found that the adverse effects of alcohol would be extinguished by about the thirteenth to fifteenth generation.

One of Stockard's papers advocated against Prohibition, entitled "Alcohol a factor in Eliminating Racial Degeneracy," was published in *The American Journal of the Medical Sciences*.[85] Dr. Charles Stockard, studied the effects of alcohol on developing fish, mice, guinea pigs and other animals. He viewed as an advantage the fact that alcohol would disproportionately eliminate the weaker and less resistant germ cells of what he termed "lower races." He felt that the American government should encourage the masses drinking alcohol since it would eliminate

---

[85] Charles R. Stockard. Alcohol a factor in Eliminating Racial Degeneracy. *The American Journal of the Medical Sciences* (April 1924, Volume 167, Issue 4; pp 469-476.

deficiencies of "inferior stock." His research in mice demonstrated that alcohol decreased the numbers of viable pups in a litter for the first several generations, which had birth defects, small pup sizes, and poor health. After the mice were exposed for thirteen or more generations, the litters were larger, more viable, and healthier – that somehow they had become resistant to the effects of alcohol.[86] The work of Stockard and other teratologists contributed to the eugenics movement promoted by the Nazi's and leading to policies of sterilization of groups such as disabled, weak and "feeble minded" individuals.

He recommended allowing alcohol to freely flow *en masse* because of its widespread developmental effects, especially on minority communities. In his 1924 seminal paper in the *American Journal of the Medical Sciences*, entitled "Alcohol a Factor in Eliminating Racial Degeneracy," Stockard described his thesis that unwanted "germ cells" causing degeneracy could be eliminated by selective exposure to alcohol. It was clear that he understood the damaging effects of alcohol on the offspring of chronic alcoholics. He believed that after several generations of prenatal alcohol exposure, the human species would be eliminated of defective genetic potential. The implications of his research today may explain why aboriginal populations are genetically more susceptible to European populations that have had prenatal alcohol exposure for hundreds of years. In fact, a type of natural selection and/or genetic tolerance to ethanol may have occurred in the heavy alcohol consuming populations of Scandinavia and Northern Europe compared with those aboriginal groups in the Americas, Australia, and Africa that have had fewer generations of exposure.

In the late 19th century, criminology research was in its infancy – led by Dr. Cesar Lombroso[87] – an Italian psychiatrist who examined living and deceased individuals in penal institutions (jails, prisons, etc.) all over Italy. He made measurements of head circumferences, faces, height, weight and conducted post mortem autopsies to provide evidence for his

---

[86] Charles R. Stockard. 1910: Influence of Alcohol and other Anaesthetics on Embryonic Development;" *American Journal of Anatomy*, Vol. 10, pp. 369-392. 1913: "Alcoholic Injuries to Germ Cells;" *American Naturalist*. 1914: "Alcoholic Injuries to Germ Cells." *Journal of Heredity*.

[87] Lombroso, Cesare, translated by Horton, Henry Pomeroy. *Crime: Its Causes and Remedies*. Published by Patterson Smith; 1911.

understanding of the biological influences on criminal behavior. He and other researchers of his era[88] began asking the question: "Do we look at the crime or the individual?"[89] in order to better understand the biological and constitutional influences of deviant social behavior. A thorough analysis of Lomboroso's methods in Havelock Ellis's translations into English are found in a recent book, *Guilty but Insane: Mind and Law in Golden Age Detective Fiction* by Samantha Walton.[90] Throughout Lomboroso's writings, there is evidence that he understood some degree of connection between alcohol and crime, "feeblemindedness" and prenatal alcohol exposure, and neurobiology and crime. His work eventually led to the biopsychosocial bases of crime – looking at the underlying mechanisms that contribute to criminal behavior. A quote from Dr. William Healy, of *The Individual Delinquent* in 1918 reads,

> *A fine example of controlled experiment is that by [Dr. Charles] Stockard, who has most cautiously studied the effect of alcohol on the germ cells of animals.*[91]

Since Caucasians have been using alcohol heavily since Biblical times, most Caucasian populations have already experienced prenatal alcohol exposure for many, many generations. In the times of the Vikings in Europe, alcohol provided internal "antifreeze" for exploitation of faraway lands in England, Ireland, and even places as distant as Iceland, Greenland and Newfoundland. English colonists used alcohol instead of water on the long journey to the Americas. These multi-generational exposures allowed for genetic resistance and tolerance in the populations, providing population-based *in vivo* support for Stockard's finding in generations of mice two centuries later. Indigenous populations – Native American, First Nations, native Africans, and people of Alaska, Australia, New Zealand and other aboriginal groups are more vulnerable because they had not endured millennia of prenatal alcohol exposure. Land-greedy European aristocrats in early America understood this concept

---

[88] The English Convict: A Statistical Study. Attributed to Havelock Ellis's *The Criminal*.
[89] Charles Goring, MD; HM Prison, Parkhurst. London 1913.
[90] Samantha Walton. *Guilty But Insane: Mind and Law in Golden Age Detective Fiction*; pp. 137-145; Oxford University Press, 2015.
[91] William Healy, *The Individual Delinquent*. 1918, page 263.

and capitalized on the vulnerability of the indigenous populations. In addition to tuberculosis blankets, the distribution of "fire water" was a known policy among traders, treaty negotiators, and others in exchange for land from Native Americans. In South Africa, pre-Apartheid policies on vineyards provided workers with ample wine to take home in lieu of adequate monetary payment, leading to couples staying inebriated from Friday evening through Sunday evening. Sobering up for work on Monday morning, they would intermittently vacillate between heavy binge exposures and abstinence. These intermittent binge amounts led to high numbers of the aboriginal children in certain communities of South Africa with heavy prenatal exposure – with rates estimated to be among the highest in the world.

Just prior to the repeal of prohibition, Aldous Huxley wrote *Brave New World* about a tiered-class society created through selective breeding en masse using the technique of *in vitro* fertilization and artificial wombs. Huxley borrowed these concepts from Stockard and other embryologists (scientists who study development), who were experimenting with nonhuman species at the time. In his novel, human eggs were fertilized in test tubes and the resulting embryos developed in carefully monitored flasks in an elaborately controlled environment. The future world leaders, Alphas and Betas, were spawned by fertilizing eggs from the tallest, most beautiful, intelligent people and grown in the ideal solution of nutrients. Other flasks of embryos were "cloned" from imperfect, less intelligent, shorter individuals, and dosed with, alcohol – the quintessential neurodevelopmental teratogen – exquisitely timed to both stunt growth and interfere with brain development. Small doses of x-rays were used intermittently to induce slight mutations – genetic changes that would ensure future generations of imperfections. These babies developed into Epsilons and Deltas, less intelligent lower caste worker drones, automatons acting at the bid and call of the Alphas and Betas of the society. Neither perfect nor defective, the middle class Gammas formed the bourgeois consumer conformists – content with mediocrity, mindless entertainment, and recreational activities distracting them from the reality of social disparities.

If learned men of the 19th and 20th century understood the neurotoxic potential of alcohol, why is it that we know so little in our own society about this prevalent and preventable condition? From my perspective, a number of scientific, philosophical, and moral principles of the 20th century led to societal amnesia about effects of prenatal alcohol exposure. First, discoveries of the early to mid-20th century led scholars to emphasize the role of genetics in development, discounting the role of environmental chemicals such as alcohol. As the field of genetics emerged, scholars reinforced a belief that criminality, pauperism, and mental problems were from inherited (genetically transmitted) conditions. Simultaneously, the eugenics movement and associated Nazi atrocities eventually led to sterilization policies for people with diminished intellectual potential, criminal behavior and mental illness to "cleanse" society of what was seen as "imperfect stock." During that era, children and adults with developmental disabilities were hidden from society, locked away in closets, attics, basements, and orphanages, or otherwise warehoused in institutions. Individuals in institutions who were deemed "feeble minded," "handicapped," or "insane" were especially targeted in eugenics campaigns. During the mid-20th century, advances in social consciousness led to reforms in child welfare. Children previously hidden away in institutions and orphanages were returned to their communities. More details of this history is discussed in a later chapter on Alternatives to Warehousing.

Finally, following the repeal of Prohibition in 1933, the earlier published scientific and public health studies showing the effects of parental alcohol abuse on infants and children were discounted as being "highly moralistic," having contributed to the Temperance movement. During World War II, many women either served in the military or worked in factories to support the war effort. The shift from homemakers to bread winners led many to drink socially, yet the hard working lifestyle left little time to party with few men home on furlough. However, the resultant Cultural Revolution of women's rights, the social pressures brought on by the Korean and Vietnam Wars, inflation, and the Cold War led to more young women drinking moderate to heavy amounts of alcohol. Unfortunately, by that time society was once again ignorant that prenatal alcohol exposure causes brain damage. The concomitant shift in cultural values and gender equality, as well as relatively looser social mores of the late 1960's and early 1970's provided a rich birthing ground for a modern epidemic of moderate to heavy prenatal alcohol exposure.

In 1973 – exactly 40 years after the repeal of prohibition – Kenneth Lyons Jones and David Smith described the Fetal Alcohol Syndrome (FAS) in the United States.[92] Five years earlier, during 1968, a French physician, Lemoine, and colleagues had identified a similar pattern of physical problems in children born to alcoholic women.[93] Despite these sentinel findings, physicians were reluctant to accept that alcohol could cause harm to a developing baby. They felt that alcohol had been consumed by populations for generations and was nearly as ubiquitous in certain cultures as water or other beverages. Obstetricians were administering "alcohol drips" (intravenous ethanol) to prevent preterm labor, casually prescribing a glass of wine to help ease pregnancy-related aches and pains or help get a better night of sleep, and encouraging breastfeeding women to "drink a beer to help their milk drop." Why should anyone believe that alcohol use during pregnancy should be discouraged?

Because of our social enmeshment with alcohol, up to 60% of pregnant women were drinking alcohol at the time of the first post-Prohibition government warning about the potential harm to a fetus. Amidst mounting evidence from animal and human studies that alcohol was a teratogen (chemical causing birth defects), in 1977 a young psychologist, Dr. Ken Warren, at the National Institutes on Alcohol Abuse and Alcoholism (NIAAA) wrote an FDA Alert advising consumers against alcohol use during pregnancy.[94] He decided on the arbitrary benchmark of a 6 drink per day limit to avoid NIAAA phone

---

[92] Jones KL and x

KennethL. Jones

Smith DW. Recognition of the Fetal Alcohol Syndrome in Early Infancy. *Lancet*; 3 November 1973; Volume 302, No. 7836, p. 999–1001. xDavidW. Smith

[93] Lemoine, Paul, H. Harousseau, J. P. Borteyru, and J. C. Menuet. "Les Enfants de Parents Alcooliques. Anomalies Observees a Propos de 127 Cas," (Children of Alcoholic Parents: Abnormalities Observed in 127 Cases) *Ouest Medical* 21(1968): 476–82.

[94] *FDA Drug Bulletin* 7(4):18, 1977.

lines being flooded with frantic callers who had been consuming in risky amounts. In order to warn the public about the mounting evidence that alcohol was a teratogen – an agent that caused birth defects very early in development, in 1981 the U.S. Surgeon General issued an advisory recommending that pregnant women and those planning pregnancy should consume no more than 2 drinks per day, but that the safest option was to abstain altogether. Dr. Warren, who drafted both the 1977 and 1981 advisories, has since confided to me that the decision was an arbitrary "non-zero" cutoff, in light of the norm to drink during pregnancy and medical practice at the time.[95]

Physicians helped perpetuate the problem, unwittingly undermining government efforts to stem the tide of the epidemic. During the mid to late 20th century, it was common medical practice to prescribe heavy doses of intravenous (i.v.) ethanol to the point of intoxication for preterm labor. Up until the early 1980s, even in large university teaching hospitals, these "alcohol drips" continued to be used unsuccessfully to prevent premature delivery.

I have met one woman, a professional living in suburban Maryland who was administered "alcohol drips" for two of her three pregnancies delivered at Columbia Hospital for Women in Washington, D.C. during the 1980s. The nurses' notes from her medical chart documented the consequences of the appalling practice: "patient giggling … slurring speech … nonsensical … falling off the gurney" as well as "nausea … vomiting … hallucinations." After several hours of the i.v. ethanol, she was sent home with a prescription to "drink at least 3 beers if contractions recur." Both her exposed children, now adults, had coordination problems and mild developmental delays, and struggled in school due to significant learning issues. Many women suffered delirium and seizures as a result of these measures, and countless babies were exposed for hours on end while the dose was being titrated. This method delivered heavy amounts of alcohol straight into the bloodstream without ever passing through the mother's liver.

---

[95] A discussion with Dr. Warren at the National Organization on Fetal Alcohol Syndrome; Wednesday, September 17, 2014; Embassy of Italy, Washington, DC.

Finally, around the time the first government advisory was issued, Dr. Ernest Abel wrote an article about the risk – benefit ratio, warning that potentially harmful effects to the fetus outweighed any limited benefit. In his abstract, he urged physicians to discontinue the practice of intravenous alcohol during pregnancy, stating that its efficacy was no better than bed rest to prevent preterm labor symptoms. He also cautioned against the practice due to harmful side effects in the mother and fetus, including death.[96] Around the same time, a drug called ritodrine replaced the alcohol drip to stop contractions. Although physicians stopped directly administering alcohol, many continued to encourage pregnant women to drink to help relax, to help "strengthen their blood," to help their milk "flow better" or "drop" during lactation. These practices have continued to some extent even today, despite nearly 35 years of multiple medical societies, government agencies, and advocacy organizations urging abstinence from alcohol at any point in pregnancy.

The 1977 FDA Bulletin and subsequent 1981 U.S. Surgeon General's Advisory cast a shadow of doubt over the safety of prenatal alcohol use. In 1981, Dr. Kathleen K. Sulik demonstrated in a mouse model that at times corresponding to as early as the third week after human conception, large doses of alcohol can cause the characteristic facial features of full Fetal Alcohol Syndrome (FAS) and associated brain damage.[97] Her research, which was published in the journal *Science*, has since been replicated in additional species and the results evidenced in human case-control and cohort studies of binge exposures prior to pregnancy recognition. Dr. Sulik's research was a tipping point that should have influenced alcohol labeling regarding early gestational exposures. Details about the need for reforms in labeling are found in a subsequent chapter on the topic.

The media hype about "cocaine babies" of the 1980s further contributed a veil of silence to public health warnings about alcohol's

---

[96] Abel EL. A critical evaluation of the obstetric use of alcohol in preterm labor. *Drug Alcohol Depend.* 1981 Jul;7(4):367-78.

[97] Sulik KK, Johnston MC, and Webb MA. Fetal alcohol syndrome: Embryogenesis in a mouse model. *Science,* 1981; 214:936-938.

harm to the fetus. As the public's perception of "crack babies" intensified, society was deafened to early pregnancy alcohol exposure as cause of birth defects and brain damage. Since cocaine is not a socially sanctioned drug like alcohol, its effects seemed more damning than alcohol. Cocaine-exposed babies experience a severe withdrawal response from the stimulant, leading to the need for special nursing care and life support in many cases. The babies born to crack addicts could be identified by urine drug tests or hair samples at delivery; whereas alcohol's relatively quick half-life made it difficult to identify at birth. Hence, amidst the growing hype about illegal substances in pregnancy, the public may have heard relatively little about alcohol in the years following the 1989 warning label.

Fairly early in the so-called "epidemic" of crack babies in the late 1980's, researchers found that cocaine is relatively tightly bound to the placenta. It became widely known that the babies were actually more affected by the alcohol their mothers consumed than by the cocaine. There were some physical birth defects in the children exposed only to cocaine, such as a missing kidney, but relatively fewer learning and behavior issues by adolescence. On the other hand, cocaethylene, a byproduct of alcohol and cocaine, metabolized in the mother's liver, crosses the placenta and blood-brain barrier in the fetus, and affects brain development more than cocaine itself. Women who use other mind-altering substances also frequently use alcohol to extend the high – so that it is not quite so high and doesn't crash too low, but is sustained longer. In that way, the alcohol and its metabolites also stay in her system longer, causing even more damage than it would otherwise.

Public perception of crack-exposed babies was likely only part of the reason for the public health catastrophe, resulting in a 400% increase in moderate to heavy alcohol use during pregnancy from 1991 to 1995.[98] Around the time of the warning label in the late 1980s and early 1990s, there was a brief upsurge in media campaigns and public health funding of prevention initiatives to heighten awareness that alcohol should be avoided during pregnancy. Alcohol labeling began in 1989, the same year that *The Broken Cord* by Michael Doris won the National Book Critics

---

[98] *Morbidity and Mortality Weekly Report*, April 21, 1997.

Circle Award. The National Organization on Fetal Alcohol Syndrome (www.nofas.org) was established in 1990 by Patti Munter in South Dakota after seeing the devastation of Fetal Alcohol Syndrome (FAS) in Native American communities. These events led to increased media exposure to the pregnancy-related risks of our social drug of choice. Less information in the public domain from fewer media messages during subsequent years led to public amnesia about drinking during pregnancy. Hence, the "out of sight out of mind" adage played out in increasing moderate to heavy alcohol use among pregnant women.

Today, social ambivalence tied to the individual rights movement and a shift in the culture of childbearing age alcohol consumers continue to perpetuate the problem of drinking during pregnancy. The Office of Juvenile Justice and Delinquency Prevention's (OJJDP's) Underage Drinking Enforcement Training Center estimates that the cost to the United States in 2010 of babies born with FASD to mothers ages 15 to 20 was more than $1.3 billion alone. A 2007 study reported in the Primary Care Companion of *the Journal of Clinical Psychiatry* by Dr. Bhuvaneswar and colleagues, "Alcohol Use During Pregnancy: Prevalence and Impact" reported rates as high as 15-20% of pregnant women consuming alcohol. The authors of the study pointed out that "downstream dysfunction" in the child may not manifest at birth but instead may lead to learning problems, pervasive developmental disabilities, behavioral issues, and intrinsic risk of alcohol and other drug abuse.[99]

Further, increases in international adoptions during the late 1990s through the first two decades of the twenty first century brought approximately 60,000 children from the former Soviet Union to the U.S. The post-cold war era social instability and poverty in Russia and Eastern Europe led to epidemic rates of alcohol abuse and ND-PAE in those countries. Hence, studies published in international journals show up to 50% of children in orphanages have been found to have ND-PAE. As a result of these demographic changes, more children with ND-PAE have been mainstreamed back into society. We are able to identify the effects of alcohol on children who are exposed to excessive amounts more

---

[99] *Primary Care Companion of the Journal of Clinical Psychiatry.* Vol. 9(6): 455–460; http://www.ncbi.nlm.nih.gov/pmc/articles/PMC2139915/

easily than those with more moderate prenatal alcohol exposure. With higher numbers of affected children to look at, we are more easily able to recognize the dark underbelly of our beloved social drug of choice.

Nonetheless, historical efforts of the government to enlighten the public about this epidemic has met with passive resistance from one of the most powerful global lobbying machines – the alcohol industry. An understanding of the history of the "Surgeon General's warning" illuminates the fact that big alcohol could do more to protect the public from harm.[100] Despite multiple ensuing Congressional hearings following post-Prohibition era recognition of brain damage associated with maternal alcohol use in 1973, it took more than a decade until 1988 for the alcohol industry to acknowledge overwhelming medical evidence that alcohol caused birth defects and brain damage. After a failed attempt for tighter government oversight and beverage labeling, the alcohol industry eventually agreed to a small label in 1988 – less than optimal for consumer protection and a far cry from the rapid response of the government toward the pharmaceutical industry in the thalidomide disaster (see earlier chapter entitled, *The Silent Epidemic*).

Not even prescription medications so profoundly benefit humanity or government economies to outweigh irreversible harm to the unborn. Alcohol should be viewed in a similar light. Yet the pharmaceutical industry of the early mid-20th century was no comparison to the megalithic lobbying agent of today's alcohol industry, with the capacity to influence government policy, shape public opinion, and influence the lifestyle behaviors through provocative and sensationalized media campaigns and marketing strategies. In a chapter of a recently published international book on the ethics of Fetal Alcohol Spectrum Disorders, my co-author, attorney Laura Riley and I argue that the alcohol industry has not been held accountable for a duty to warn – which is a failure in consumer protection. By not enlightening the public about the scientifically proven fact that its product can harm children's brain development before a woman knows she is pregnant, the alcohol industry

---

[100] Rich SD and Riley LJ. Neurodevelopmental Disorder associated with Prenatal Alcohol Exposure: Consumer Protection and the Industry's Duty to Warn. *Fetal Alcohol Spectrum Disorders: Ethical and Legal Perspectives.* Springer Publications, Amsterdam, 2015.

is liable for damages incurred to babies unintentionally exposed to the noxious substance prior to pregnancy recognition.

Alcohol is a socially-sanctioned substance that a majority of the world's population uses liberally and with very little regulation other than age restrictions and operating vehicles while under the influence. The spirits industry learned from mis-steps of big tobacco in the 1970s and 1980s that led to their downfall as Congressional hearings uncovered memos and company documents acknowledging public health and medical reports about the human cost of cigarettes and related products. The alcohol industry's trump card for consumer protection has been the minuscule warning label on alcohol beverages.

It was only in 2005 that a revised advisory warned against any use of alcohol during pregnancy, recommending that women who are pregnant or planning a pregnancy abstain altogether. The tiny warning label on bottles is a little bit of information conveyed too late – like arm chair quarterbacking after Monday night's big game. While the initial 1981 warning included research linking early prenatal alcohol exposure to Fetal Alcohol Syndrome (FAS),[101] the alcohol industry continues to omit a key point in its label: by the time most women know they are pregnant, much damage may have already occurred.

Ethically, the alcohol industry should be held accountable to inform consumers to avoid drinking when pregnant or planning a pregnancy, as well as to avoid pregnancy (i.e., use contraception or abstain from intercourse) if using alcohol regularly. While the alcohol industry's warning label focuses on pregnant women, a more accurate prevention strategy would use upstream approaches to teach the public to (a) plan pregnancies, (b) avoid pregnancy (i.e., contracept) if using alcohol, and (c) avoid alcohol if pregnant or planning to be.

Hopefully, *Preconceptional Perspectives* will help move the conversation from "Oops! I'm pregnant ... I'd better not drink!" to a point upstream before conception – pre-empting firewater's damage to the unborn yet to be.

---

[101] Sulik KK, Johnston MC, Webb MA. *Science*. Fetal alcohol syndrome: embryogenesis in a mouse model. 1981 Nov 20;214(4523):936-8.

Chapter 8

PRECONCEPTION HEALTH - A LOT HAPPENS BEFORE PREGNANCY RECOGNITION

Alcohol is the quintessential neurodevelopmental teratogen.

— Edward P. Riley, PhD; San Diego State University

I have often remarked that the term "Fetal Alcohol Spectrum Disorder" is a misnomer since alcohol impacts the underappreciated *embryonic* stage not just the *fetal* period. In my view, it is not enough to warn women about using alcohol after pregnancy recognition or even if they are planning to be pregnant. Women who drink in binge amounts often have irregular periods. Given that 50% of pregnancies are unplanned and 13% of women binge drink during their reproductive years, many children are unintentionally exposed to alcohol during critical early stages of pregnancy – before a woman may have realize she's missed a couple of periods.

How amazing and precarious that the most critical stages of human development occur often before we even know our future children exist. For any given child, it is a miracle that they make it beyond the embryonic period. In this precariously sensitive stage before we know they exist, our children are laying the groundwork for every organ, system, nerve, and process in their future body. The choices that we make during those critical stages will determine the fate of how their genetic blueprint plays out. While many environmental chemicals are less under our control than we would like to imagine – as eloquently documented in Rachel Carson's monumental "whistle blowing" manuscript, *Silent Spring* – in our modern society, we have the means to control if and when we become pregnant and other potentially devastating lifestyle decisions such as how much, how often, when, and with whom we are drinking or using other drugs of abuse. In order to understand why "prevention" efforts focused on pregnancy are a little too late, one must learn about the process of early development (embryology) and the study of how birth defects happen (teratology). Let's begin by immersing ourselves in the miracle of life that occurs during the first several weeks after conception.

The first eight weeks after conception are considered the embryonic stage (meaning having to do with embryo development), as well as the period of "organogenesis"[102] (related to organ development) because the major organ systems, brain, and spinal cord are forming during this time. Scientists who study early life at these formative stages are called "embryologists." Those that study the effects of drugs, chemicals and other substances on this early developmental period are "teratologists." Teratology is the branch of science concerned with the production, development, anatomy and classification of malformed fetuses.[103] [The word teratogen—meaning a drug or other agent that causes birth defects—originates from the Greek root *teras* meaning monster.[104]]

Within the first week, the rapidly dividing fertilized egg makes its way down the fallopian tube and forms a needle-point ball of cells that implants in the uterine wall. Following "implantation," she begins to form her nutritional dependence on the maternal circulation. During week two, a cascade of perfectly-timed events organizes cells of the three germ layers—precursors for the entire human architecture—and mobilize some to migrate toward their designated launching pads where they will continue their developmental journey.

---

[102] "Organogenesis" is also known as "embryogenesis," or development of the embryo. Fetal development begins in the early 9th week after conception, at which time further development and growth of the embryo continue to occur.
[103] Sadler TW, *Langman's Medical Embryology*, 6th Edition, Williams and Wilkins, Baltimore, 1990, p. 149.
[104] *The American Heritage Dictionary*, 2nd College Edition, Houghton Mifflin Company, Boston, 1982.

Susan D. Rich MD MPH DFAPA

From a sphere to a disc to a kidney bean, the embryo—suspended between yolk sac and amnion—flips and folds upon itself as organ development begins. It is late in the second week after the egg is fertilized that this process known as "organogenesis" begins. Embryologist Lewis Wolpert once said, *"It is not birth, marriage, or death, but gastrulation, which is truly the most important [vulnerable] time in your life" (1986).* Brain development, also known as "neurulation," ensues with birth of tiny cells that form in an area known as the primitive streak. An eventual failure to complete this process ends in *spina bifida* (open spinal cord) and/or anencephaly (open brain) – two common neural tube defects that can be caused by exposure to alcohol and other chemicals during this critical stage.

By the end of week three, we are merely the size of Lincoln's bow tie on the US penny (see picture) or Roosevelt's ear on the dime. Our primitive brain is precariously open and vulnerable to alcohol and other chemicals – seeping through the walls of the uterus into the yolk sac. Our baby neurons continue to migrate out toward their destinations. From this point on, millions of precisely-placed cells connect the central nervous system to the periphery through a complex web of dendrites and synapses – like a complex neural circuitry of a computer's mother board or a modern voice-activated home. These cells of our human computer are destined to release neurotransmitters in response to mathematical problems, stressful stimuli, thought processes, and other situations of the human condition.

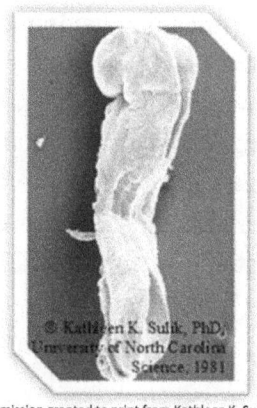

Permission granted to print from Kathleen K. Sulik, PhD

We are the size of Lincoln's bowtie on the US penny at a most vulnerable time in neural development.

The two sides of the face begin to mesh as the hollow neural tube—the primitive spinal cord and brain—zips up, forever hiding the miswired neurons within layers that continue to form, eventually covered by a face and skull that only echo subtle clues to the damage within. Early in week four, microscopic cells of the rudimentary pacemaker spontaneously begin to beat, the first pulsations that will send blood racing through the highways of arteries and veins yet to be laid down. The first missed menstrual period may or may not be noticed (Figure 4) – leaving the developing embryo further at risk of harm from her mother's unwitting lifestyle behaviors. By week five, four tiny limb buds erupt into flippers, two on each side of the midline. The arms sprout first and the legs follow—destined to run through forests, across school yards, up mountains, or fleeing a crime scene.

Beginning in week six, fingers blossom from paddle-like hands that one day will pick noses and flowers. By the end of week seven, the primordial gills—branchial arches—mold into ears and the tiny undifferentiated genitals appear. From this stage, gender-specific characteristics develop in response to the exquisitely timed release of hormones that influence development of our sex organs. Newly formed eyes—fated to peer into a microscope or down the barrel of a gun—are last continuously exposed to the dimly lit, watery abode during the eighth week. Lids will seal them off in the ninth week until they are reopened months later for future batting of eyelashes or winking at a prospective mate.

The field of "teratology" (study of birth defects) reveals the exquisite vulnerability of these intricate physical and molecular processes. What comes as no surprise to the budding developmental scientist is that numerous environmental chemicals disrupt these synchronous mutually dependent events. Such toxins that cause permanent disfigurement (physical birth defects) are known as *teratogens*.[105] Perhaps the historically most notorious physical teratogen, thalidomide disrupts the process of long bone development during weeks five through seven, leading to babies being born with foreshortened limbs. Our delicate

---

[105] A *teratogen* is a chemical that causes birth defects. Physical teratogens cause damage to structures such as limbs, organs, eyes, ears, palate, lips, etc. Functional teratogens cause damage to neurons in the brain, spinal cord, and peripheral nervous system.

brain is sensitive particularly to chemicals that are "lipid-soluble," that is, they dissolve in fats, such as solvents. Cell membranes are made of lipid bi-layers, so solvents and inhalants that are "fat loving" or "lipophilic" cross barriers like the blood vessels of the placenta and the brain. These neurotoxic chemicals are known as *neuroteratogens*, causing injury to the brain and nerves during development. Unseen at birth, the resulting neurologic, mental and intellectual disabilities are known as "functional" (rather than physical) birth defects.

The Food and Drug Administration (FDA) has established six classes to indicate a drug's potential to cause birth defects if used during pregnancy. The classes are based on a combination of documented evidence of the drug causing birth defects and its "risk-to-benefit" ratio (i.e., how much value it has for treating a condition relative to its potential for harm to a developing child). The six classes are listed in alphabetical order, A through D and X.[106] "Category X" drugs are known to cause physical birth defects at high rates and the benefit must outweigh the risk to the consumer. The FDA updated its prescription drug labeling effective June 30, 2015 as follows:[107]

8.1 Pregnancy

Pregnancy Exposure Registry

> If there is a scientifically acceptable pregnancy exposure registry for the drug, the following statement must appear:
>
> *"There is a pregnancy exposure registry that monitors pregnancy outcomes in women exposed to (name of drug) during pregnancy."*

The statement must be followed by contact information needed to enroll in or to obtain information about the registry.

---

[106] http://www.drugs.com/pregnancy-categories.html
[107] http://www.fda.gov/Drugs/DevelopmentApprovalProcess/DevelopmentResources/Labeling/ucm425415.htm

Risk Summary

Provides 'risk statements' that describe for the drug the risk of adverse developmental outcomes based on all relevant human, data, animal data and the drug's pharmacology.

When applicable, risk statements must include a cross-reference to additional details in the relevant portion of the Data subheading in the Pregnancy subsection.

Clinical Considerations

Provides information to further inform prescribing and risk-benefit counseling. Relevant information is presented under the following 5 subheadings:

- *Disease-associated maternal and/or embryo/fetal risk*
- *Dose adjustments during pregnancy and the postpartum period*
- *Maternal adverse reactions*
- *Fetal/Neonatal adverse reactions*
- *Labor or delivery*

Data

Describes the data that provide the scientific basis for the information presented in the Risk Summary and Clinical Considerations

- *Human Data*
- *Animal Data*

8.2 Lactation Replaces the Nursing Mothers subsection

Risk Summary

Summarizes information on the presence of a drug and/or its active metabolite(s) in human milk, the effects of a drug and/or its active metabolite(s) on the breastfed child, and the effects of a drug and/or its active metabolite(s) on milk production

Clinical Considerations

Provides information to further inform prescribing and risk-benefit counseling.

- Minimizing exposure
- Monitoring for adverse reactions

Data

Describes the data that provide the scientific basis for the information presented in the Risk Summary and Clinical Considerations

8.3 Females and Males of Reproductive Potential

Includes information for these populations when:

- There are recommendations or requirements for pregnancy testing and/or contraception before, during, or after drug therapy; and/or
- There are human and/or animal data suggesting drug-associated effects on fertility

We are enlightened enough as a society to protect the unborn and the unborn yet to be from pharmaceutical medications and insist that manufacturers accurately label their products, report known adverse events, and protect the consumer by alerting the public to potential side effects or risk of harm. Even those with life-saving properties such as cancer drugs, HIV/AIDS treatment, and other health benefits are not immune to the rigors of the FDA's concerned oversight to guard against harm – even prior to pregnancy recognition. Yet prenatal alcohol exposure can be much more damaging to the developing brain than many medications and illegal substances. Why do we as a society have such a difficult time appreciating that a substance so widely used and seemingly innocuous as alcohol could predispose individuals to such far-reaching social consequences as learning disabilities, deviant behaviors, employability issues, and homelessness? Endemic (low baseline level) rates have existed since man first fermented berries and fruits. As discussed in the chapter on Historical Perspectives, epidemics (outbreaks) have occurred when social systems fail, such as in times of political unrest, financial collapse, or other social hardship.

How does ND-PAE Happen?

It's been 35 years since Dr. Kathleen Sulik's work was first published during 1981 in the prestigious journal *Science* showing that alcohol causes brain damage as early as the third week after conception. Yet our society has overlooked the ugly secret of how early ND-PAE happens. Ironically, acting as both a libidinal lubricant (disinhibiting of sexual impulses) and a "neurodevelopmental teratogen," alcohol is a clandestine neurotoxin that kills brain cells early in development. Like a virus or parasite that perpetuates itself inside an unwitting host, alcohol's covert maneuvers allow damage by exchange through the sacred body waters of mother to offspring. Alcohol is perhaps the most widely accessible (ubiquitous) neuroteratogen, silently crossing the placenta and entering the fluid filled space where the embryo develops. The ninja solvent and its metabolite, acetaldehyde disrupt brain and nerve development by killing migrating cells along their pathway. In this way, alcohol's damage to the child is mostly invisible to the naked eye, causing more harm to the developing brain than to other systems.

The elegance of Dr. Sulik's research is in the simple common sense behind the implications of her scientific findings. In 1981, she and her colleagues showed that the equivalent of four to five servings of ethanol exquisitely timed in the late third week after conception was enough to cause programmed cell death, or apoptosis, in important midline structures of the embryo. She taught pregnant mouse dams to drink the equivalent of one binge episode during what would be the late third week in the human. She harvested the pups two hours later and stained them with Nile blue stain, which is taken up by dead cells. Areas of apoptosis (cell death) are noted in the slide shown but not seen in the control pups' brains. The cells that are most sensitive are destined to become the midline areas of the face and brain. Since the facial defects and midline problems of the brain occur during the

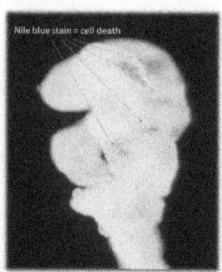

Permission granted to print from Kathleen K. Sulik, PhD; LE Kotch and KK Sulik, Int. J. Devl. Neuroscience 10: 273-279, 1992

embryonic period, the term "alcohol embryopathy" more accurately describes the disorder rather than Fetal Alcohol Syndrome, which implies fetal damage later than the ninth week after conception. Nonetheless, there could be no better designed way to cause mental deficits in offspring than to have a chemical that disinhibits behavior (i.e., causing promiscuous sexual behavior), leading to intercourse, and surreptitiously causing brain damage before pregnancy is recognized.

Recent research has taught us that developing neurons exposed to alcohol leaves them like a ship on the open ocean, stranded without a compass. In this way, alcohol disrupts neuronal signaling, preventing cell migration to their proper destination. Like ants who lose the chemical "scent" of the scout on the way to their food, alcohol disrupts these baby brain cells making their way along the path to their destination in the brain by interacting with a cell adhesion molecule, called LCAM. Just as tree limbs may die if pruned in springtime before blooming, alcohol can destroy brain cells that are migrating or developing, snuffing out their life. Alcohol can also affect the myelin sheath around neurons so that neuronal conduction is less efficient and sometimes triggers firing of neighboring neurons. The resulting invisible neuroanatomical defects are as devastating as physical deformities—and as preventable.

Prenatal alcohol exposure can also cause a diminished basal ganglion, a multi-lobed structure deep in the brain, controlling and extending to other areas that affect cognition, emotion learning, voluntary motor movements, and routine habits such as bruxism (grinding teeth) and eye movements. Important extensions of the brain known as cranial nerves can also be affected, including those controlling the muscles of the head, neck and face. Specific lobes of the cerebellum can also be depleted. These and other parts of the brain direct mental processing, abstract thinking, reasoning, emotional control, and fine movements—functional birth defects which cannot be seen with the naked eye. Hence, these functional birth defects lead to emotional dysregulation, anxiety, neurocognitive and learning issues, social communication issues, and sensory/motor problems. Therein lies the term neurodevelopmental deficits.

Ironically, these cells will be the precursors to important mental and involuntary nervous system functions like working memory, processing speed, problem solving, arousal, mood regulation, language and speech, sociability, gullibility, impulse regulation, and consequential thinking. Deficits in these areas can lead to predisposition to deviant behavior without proper social and environmental supports as children transition to adulthood. Neurologists describe the brain damage as "static encephalopathy," which ranges from static (i.e., not causing changes on encephalography or EEG) to epileptiform (i.e., causing changes on EEG) and epileptic (i.e, causing overt seizure activity). Sadly, without proper guidance and supervision, these young adults become marginalized in a society that favors social intelligence, strategic thinking, and self-control.

How much is Too Much?

Alcohol affects women and offspring differently based on a number of factors. Not everyone is equally effected by prenatal alcohol exposure and even moderate to heavy exposures can lead to only subtle problems, such as learning and attention issues in children. First, females have half the alcohol dehydrogenase enzyme in their stomachs to digest alcohol. Therefore, their blood alcohol concentrations will rise more quickly than their male counterparts. Other variables include a person's genetics, metabolism, nutrition, age, as well as the timing, duration, frequency of exposure. For example, one woman may be prone to a higher blood alcohol level because she is from an ethnic group, such as certain Asian people, that have a genetic variant in the enzyme alcohol dehydrogenase that digests alcohol. Another woman may be at risk of a higher concentration of alcohol in her system because she has liver problems due to chronic alcohol abuse. A third may be vulnerable because she has a higher tolerance to alcohol and doesn't feel its full effects until she has blacked out. Others may have poor nutrition or may be smokers – either leaving their developing child without important protective factors of the right balance of nutrients and oxygen to help prevent some of alcohol's damaging effects.

In our modern era, we have understood that babies born to women as they progress through their lifecycle of alcoholism (from "social drinking" to "binge drinking" in their younger reproductive years to daily heavy drinking habits) as well a progression in their liver dysfunction and nutritional status as they age. Anne Streissguth at the University of Washington at Seattle and many other researchers has shown increasing age as a risk factor of worse outcomes for children born to those mothers. At the same time, children who are born to women who are early in their reproductive lives – binge drinkers who have unintended or mistimed pregnancies for their first pregnancy then plan future pregnancies (avoiding alcohol before becoming pregnant) would have worse outcomes in their older children.

A young woman, Celeste, shared with me recently that a friend became pregnant during an extended episode of post-college partying and socializing. By the time the young woman learned she was pregnant, several weeks had passed since conception. Believing that a little alcohol would be okay during pregnancy, she continued to drink – albeit at a lower rate than she had been earlier in the pregnancy. The baby was born with two eyes, a nose, mouth, ten fingers and ten toes, as well as other body parts seeming to be well formed and in the right places. Celeste's friend breathed a huge sigh of relief that her earlier drinking had no apparent effect on the baby. While her child may grow up to have no significant effects of her drinking, hopefully she will recall her alcohol use during the stages prior to pregnancy recognition should her child have learning issues, attention problems, emotional outbursts, or other neuropsychiatric issues.

Certain mothers and fetuses are at greater risk, such as women of Asian descent who often have a genetic variant of alcohol dehydrogenase in their stomach, allowing more rapid increases in blood alcohol concentrations. As an example, a very well educated Asian mother. Linda, brought her 15-year-old son, Kevin, to see me for evaluation by recommendation of a neuropsychologist who suspected ND-PAE. She and Kevin's father had met while they were in graduate school. They were each drinking two drinks of wine a day at that time, in line with her understanding of the American Heart Association (AHA) guidelines.[108] To clarify for the reader, the AHA website states the following:

> *If you drink alcohol, do so in moderation. This means an average of one to two drinks per day for men and one drink per day for women. (A drink is one 12 oz. beer, 4 oz. of wine, 1.5 oz. of 80-proof spirits, or 1 oz. of 100-proof spirits). Drinking more alcohol increases such dangers as alcoholism, high blood pressure, obesity, stroke, breast cancer, suicide and accidents.*

Since the AHA website did not mention at that time anything about the dangers of drinking while pregnant or planning pregnancy (or using birth control if following their recommendation), Linda reported drinking two drinks a day on average – one drink more than recommended by the AHA. Since most people under report their drinking, there were likely times that Linda drank more, such as when they were out with friends socially or celebrating a holiday or special occasion. Nonetheless, she immediately stopped drinking at 9 weeks when she learned she was pregnant. Over the intervening years, doctors had only ever asked whether she drank during the pregnancy, not if there might have been exposure prior to pregnancy recognition. Two key questions missing from most child psychiatrist's history form are: *Was this a planned or mistimed pregnancy? Might there have been alcohol use prior to pregnancy recognition?* Linda went on to finish her graduate program and began working in research. Kevin, now in high school, struggles with social communication problems, mood dysregulation, executive functioning issues, and a variety of motor and sensory problems. It was only an astute neuropsychologist who understood the "red flags" to look for on neuropsychological testing – gaps and inconsistencies in test results – some remarkably average or above average scores with borderline to very low scores in other areas.

For Kevin, he suffered from extreme difficulties in working memory and auditory integration, while his processing speed and verbal

---

[108] Alcohol and Heart Health, American Heart Association guidelines. http://www.heart.org/HEARTORG/GettingHealthy/NutritionCenter/HealthyEating/Alcohol-and-Heart-Health_UCM_305173_Article.jsp.

comprehension scores were average. He also had deficits in speech pragmatics, not understanding the back and forth of communication, the subtle nuances of nonverbal cues, and the rhythm of conversation. While his overall IQ was about 95, the deficits in many areas of executive functioning and social communication placed him squarely in the 0.3 to 2% range, meaning 98 to 99.7% of people scored better than him.

Now one might ask what it is about Linda that made her and/or Kevin susceptible to the effects of alcohol on pregnancy. Although she maintained a healthy diet, regular exercise, and had no other medical problems, she and other people of Asian ancestry carry a genetic variant in their alcohol dehydrogenase enzyme – the metabolic enzyme in the stomach that helps digest alcohol. Because of the difference in their clearance of alcohol, their blood alcohol concentration reaches a higher level more quickly than non-Asians, leading to a "flush" response. It is that response that makes many people avoid drinking. Because of the higher concentration of alcohol in Linda's system, Kevin was exposed to a higher amount over a longer period of time. The amniotic fluid collects the alcohol, which takes much longer to clear from the fetal circulation than from the mother's because of the lack of development of the stomach and liver enzymes. Unfortunately, the developing embryo and fetus is unable to adequately metabolize alcohol, so the concentration builds within the fluid filled home at a higher concentration than in the mother's blood.

This brain damage associated particularly with early, binge, or frequent prenatal exposure to alcohol is particularly harmful. While drinking at any point in pregnancy plays Russian roulette with brain development, the third week after conception is critical for neural crest cells (baby neurons) being born and migrating to areas they will flourish. Under the stealth power of pre-pregnancy recognition, the neuronal Internet is permanently damaged and no one will forever know. After all, when asked whether she used alcohol during the pregnancy, the mother will accurately report "no;" because she did not know she was pregnant until well into the second month.

A recent CDC Morbidity and Mortality Weekly recognized binge drinking as a "serious, under-recognized problem among women and girls,"[109] with nearly fourteen million US women binge drinking. Women who binge drink do so frequently – about three times a month – and have about six drinks per binge. [Binge drinking for women is defined as consuming four or more alcohol drinks (beer, wine, or liquor) on an occasion (i.e., over the course of an evening at home or socializing with others)]. Even more alarming is the age disparity in the drinking patterns. While one in eight women aged eighteen years and older report binge drinking, one in five high school girls binge drink. According to the report,

> *Binge drinking is a dangerous behavior but is not widely recognized as a women's health problem. Drinking too much – including binge drinking - results in about 23,000 deaths in women and girls each year. Binge drinking increases the chances of breast cancer, heart disease, sexually transmitted diseases, unintended pregnancy, and many other health problems. Drinking during pregnancy can lead to sudden infant death syndrome (SIDS) and Fetal Alcohol Spectrum Disorders (FASD).*
>
> **Drinking among Women Age 15 to 44**: *In the United States: 1 in 2 reports any alcohol use in the past month. Approximately 1 in 4 reports binge drinking (defined as 5 or more drinks on one occasion). About 1 in 20 reports heavy alcohol use (defined as binge drinking on at least 5 days in the last month).*
>
> **"Drinking Among Pregnant Women in the United States**: *in 30 pregnant women reports high-risk drinking (defined as 7 or more drinks per week, or 5 or more drinks on any one occasion). 1 in 9 pregnant women binge drinks in the first trimester. 1 in 30 pregnant women drinks at levels shown to increase the risk of FASD. More than 1 in 5 pregnant women report alcohol use in the first trimester, 1 in 14 in the second trimester, and 1 in 20 in the third trimester.*

---

[109] Centers for Disease Control and Prevention. *Morbidity and Mortality Weekly, Vital Signs:* Binge Drinking Among Women and High School Girls—United States, 2011; January 11, 2013: 62(01);9-13.

With high rates of binge drinking during reproductive years, it is helpful to consider unintended pregnancy rates in the U.S. From a 2011 article in the journal *Contraception,* the rate of unwanted or mistimed pregnancies for women aged 15-44 has not improved since the early 1980s when widespread use of contraceptives led to a decline from 59 to 49 unintended pregnancies per 1,000 women. In 1994, 49% of pregnancies were reported to be unintended, with 54% of those ending in abortion. Rates remained relatively similar in 2001 (48%) and 2006 (49%).[110,111]

A study published in 2014 examined the effects of pregnancy recognition on alcohol use among about 1000 women with unwanted (mistimed or unplanned) pregnancies in a total of 30 pregnancy termination facilities in the U.S. A majority of the sample (56%) of the women reported alcohol use in the month before learning they were pregnant. Approximately half of those reporting any use (23%) used a minimum of 6 drinks on at least one occasion. The study also looked at the numbers of women who quit drinking after pregnancy recognition and found that many did quit drinking no matter whether or not they had chosen to terminate the pregnancy.[112]

Heavy amounts of alcohol consumed in the first several weeks after conception can cause eye deformities, facial clefts, minor bone and joint deformities, as well as a number of other physical birth defects of the genitals, spine, and organs such as the heart. The term "alcohol-related birth defects" has been used to describe these anomalies, which are infrequent and less dramatic as a missing limb. Alcohol exposure timed very early in pregnancy can cause *spina bifida,* cerebral palsy, anencephaly (being born without a brain), and even cyclopea (being born with only one eye). While these birth defects are horrific, they are typically not ever related back to the mother's use of alcohol because they occur before she even knew she was pregnant. If a doctor asked her, *"Did you use alcohol when you were pregnant?"* the answer would be

---

[110] *Contraception.* 2011, 84(5):478–485.
[111] Finer LB and Zolna MR, Unintended pregnancy in the United States: incidence and disparities, 2006.
[112] Roberts SCM, Wilsnack SC, Foster DG, and Delucchi KL. Alcohol Use Before and During Unwanted Pregnancy. *Alcoholism: Clinical and Experimental Research;* Volume 38, Issue 11, pages 2844–2852, November 2014.

emphatically *"No! Of course not!"* in most cases, because she stopped after pregnancy recognition. Such severe birth defects due to alcohol are also rare requiring high blood alcohol levels which can be achieved by either very large doses of alcohol – consumed by regular binge drinkers or alcoholic women – or the simultaneous use of prescription or illegal drugs, genetic risk factors, nutritional status, metabolism, and/or other underlying physiologic issues.

Government agencies and advocacy organizations estimate that 40,000 babies are born each year with some degree of this preventable form of developmental disability. Recent large scale epidemiology studies by the Centers for Disease Control and Prevention estimate that as many as 2-5% (1 in 20) school aged children have some degree of this condition, with much higher estimates in certain ethnic groups. Horrifically, some women continue this neurotoxin throughout the pregnancy, believing erroneously that they or their baby are immune to its effects, being complacent or ambivalent about the outcomes, or ill-advised by the medical community, the media and other social systems.

While some women and/or their babies may have genetic factors that protect them from alcohol and other neurotoxins, it is impossible at this point to determine who is genetically immune to alcohol's effects. These effects can occur as early as the first few weeks after conception – before most women know they are pregnant. The central nervous system (CNS) and brain are the most sensitive and vulnerable structures to the effects of alcohol and can be affected by moderate to heavy alcohol use at any point in gestation. Further, it is clear that early exposures to four to five drinks is enough to cause full Fetal Alcohol Syndrome (FAS) – even if a person is unaware of her pregnancy. Hence, both science and the U.S. Surgeon General tells us to avoid alcohol if you are pregnant or planning pregnancy and to avoid pregnancy if using alcohol. The initial Surgeon General's advisory against alcohol use during pregnancy announced in 1981 did not receive much publicity until 1989 when labeling became a reality in the U.S.[113] Since there is no safe amount of alcohol at any point during pregnancy, all childbearing age women

---

[113] Rich SD and Riley LJ. Neurodevelopmental Disorder Associated with Prenatal Alcohol Exposure: Consumer Protection and the Industry's Duty to Warn. Chapter 3 in *Fetal Alcohol Spectrum*

should avoid alcohol if there is a potential for pregnancy. That is, if you are sexually active and using alcohol, either contracept or don't have sex.

Initially, the labeling of alcohol caused a decrease in alcohol use in pregnancy; however, studies show that alcohol use among pregnant women rose by 400% from 1991 to 1995. What this means is that fourfold more women were using moderate to heavy amounts of alcohol during pregnancy in 1995 compared with the few years following alcohol labeling. While it is unclear why more women were drinking in subsequent years following alcohol labeling, one reason may be that social perceptions of the harm in drinking diminished after initial media awareness tapered off – "out of sight, out of mind." By the early 1990s, intensity in marketing of alcohol to childbearing age women increased the numbers of "girlie drinks" massed produced that gave the illusion of less alcohol content due to their fruity beverage names and flavors. Another reason may be that prenatal alcohol-related problems are hidden or silent neurodevelopmental issues – e.g., cognitive impairment, heightened arousal (fight or flight response), social communication deficits, and sensory/motor issues, which are all less tangible than physical disabilities. So it is harder for people to accept and therefore embrace prevention strategies and to keep dangers associated with prenatal alcohol use in the collective (social) conscious.

Preconception Health: Preventing ND-PAE Prior to Pregnancy Recognition

The idea of preconception care was birthed out of a collaboration between Dr. Robert Cefalo and Merry-K Moos of the University of North Carolina.[114] Their numerous papers, books, and lectures point to reproductive health screening, intervention, and education beginning prior to pregnancy since a majority of physical birth defects and many functional (or neurodevelopmental) birth defects occur prior to pregnancy recognition. Efforts to promote use of multivitamins,

---

Disorders in Adults: Ethical and Legal Perspectives - An overview on FASD for professionals. Nelson M and Trussler M (Eds.). Int. Library Ethics, Law Volume Number:63. Springer Publications, 2015.

[114] Cefalo RC and Moos MK. *Preconceptional Health Care: A practical guide.* 2nd ed., Mosby, St. Louis, 1995.

eliminate potential harmful chemicals, emphasize regular exercise, and educate about nutritious foods and those to avoid are more helpful in the period of time when women are planning their pregnancies rather than after they are already pregnant.

Currently, the Centers for Disease Control and Prevention has a Division of Preconception Health and Health Care that is attempting to develop programs and raise awareness about healthy lifestyles before, not just during pregnancy. According to their website:[115]

> **Preconception health** *refers to the health of women and men during their reproductive years, which are the years they can have a child. It focuses on taking steps now to protect the health of a baby they might have sometime in the future.*
>
> *However, all women and men can benefit from preconception health, whether or not they plan to have a baby one day. This is because part of preconception health is about people getting and staying healthy overall, throughout their lives. In addition, no one expects an unplanned pregnancy. But it happens often. In fact, about half of all pregnancies in the United States are not planned.*
>
> **Preconception health care** *is the medical care a woman or man receives from the doctor or other health professionals that focuses on the parts of health that have been shown to increase the chance of having a healthy baby.*
>
> *Preconception health care is different for every person, depending on his or her unique needs. Based on a person's individual health, the doctor or other health care professional will suggest a course of treatment or follow-up care as needed. If your health care provider has not talked with you about this type of care—ask about it!*

Dr. Louise Floyd has been at the forefront of this effort to bridge gaps in understanding among policy makers, health care professionals, and the public in her leadership roles at the CDC. These efforts are geared to enlighten the masses about preconception care in an effort to

---

[115] http://www.cdc.gov/preconception/overview.html

prevent neural tube defects, reduce other malformations, and improve reproductive health outcomes. They and other health agencies advocate for use of prenatal vitamins that contain micronutrients found to be neuroprotective (e.g., preventing neural tube defects) and iron to prevent anemia, regular exercise to improve cardiovascular health and stamina for the pregnancy and post-partum period, well-balanced meals with fresh fruits and vegetables to improve metabolism and reduce the tendency to put on more weight than recommended during the pregnancy.

These efforts will eventually spawn a new era in women's reproductive health – including preconception planning and contraception for alcohol consumers. Preconception health messages promoted along with condoms, effectively used for prevention of HIV/AIDS and other sexually transmitted diseases, should be delivered by every agent with a duty to warn: medical providers, media, government agencies, and most of all alcohol manufacturers, importers, distributors, and retailers. Reproductive education (i.e., sex education), preconception health awareness, and increased access to contraceptives combined with knowledge that alcohol and other chemicals affect our future children before we know they exist are all important ways to prevent ND-PAE. Since the days of thalidomide, preconceptional approaches have been useful for preventing inadvertent prenatal exposure to Category X medications. Doctors routinely mandate pregnancy testing and contraceptive use when treating of a variety of medical issues, ranging from life threatening illness like cancer, to mental problems such as bipolar disorder or schizophrenia, to purely aesthetic dermatologic conditions like acne and wrinkles. In order to adequately protect a potential pregnancy from harm, Category X medications are only prescribed to childbearing age women whose medical condition is unable to be treated with alternative methods. In those cases, the patient and his/her partner will use at least two reliable forms of contraception. For example, derivatives of Vitamin A (also known as retinoic acid) that are used in prescription drugs and face creams to treat severe acne and wrinkling can cause severe malformations, intellectual impairments, and a variety of other birth defects. These products—i.e., Retin-A and Accutane—are carefully labeled by pharmaceutical manufacturers to inform consumers to avoid pregnancy. Prescribing physicians are ethically obligated to educate women of childbearing potential about the potential for birth defects, encouraging them and their partners to use contraceptives.

By shifting the paradigm in our thinking about reproductive health and family planning toward primary prevention strategies, we can begin educating young people about the links between ND-PAE, unintended pregnancies, and binge drinking. One approach is by entering the classroom discussions about reproductive health (i.e., "family life") in fifth through 12th grades and liven up the discussion with information about ND-PAE. Dr. David Satcher, the former U.S. Surgeon General, has been a proponent of increased understanding of adolescents about the importance of contraception in improving lifestyles, reducing poverty, and preventing sexually transmitted infections (STIs) as well as Human Immunodeficiency Virus (HIV).[116]

More recently, Dr. Satcher and his colleagues have described other benefits of sexual health. In an article in the *Journal of the American Medical Association*, they advocate for integration of a sexual health framework into primary care practice by incorporating health-related discussions about sexuality into various aspects of medical practice.[117] He suggests "A Sexual Health Framework" to be broadly integrated throughout the health impact pyramid, using 4 key principles to support, streamline, and enhance existing disease control and prevention activities: 1) Emphasis on wellness. 2) Focus on positive and respectful relationships. 3) Acknowledgment of sexual health as an element of overall health. 4) An integrated approach to prevention.

Missing from the debate about the pros and cons of sex education to prevent STDs/HIV and improve educational opportunities for women is that dirty secret our society is reluctant to inoculate its future citizens against – that alcohol use prior to knowledge that a young woman is pregnant can permanently damage the brain of her future child. In

---

[116] Schemo DJ; Surgeon General's Report Calls for Sex Education Beyond Abstinence; *New York Times;* June 29, 2001.

[117] Satcher D; Hook EW; Coleman E. Sexual Health in America: Improving Patient Care and Public Health; Viewpoint; *Journal of the American Medical Association.* June 18, 2015

my view, anyone who recognizes the importance of human life like Native and aboriginal people, understands that all life is sacred. Staunch conservatives like my very southern father would not argue that the minds of our future children are important enough to protect from the reckless abandon of adolescence. Perhaps they would embrace family planning approaches and sex education more from the perspective that we can make our society healthier, smarter, and higher functioning by eliminating the leading preventable cause of intellectual disability and neurodevelopmental disorders (ND-PAE) by improving knowledge about how early ND-PAE happens and providing access to reliable contraceptives.

Though there are numerous prevention campaign comparisons, the practice of public health dentistry to prevent cavities in teeth is a similar paradigm shift that will be needed to prevent ND-PAE. The solution to preventing prenatal alcohol-induced cavities in the brain will be tantamount to dentists who began teaching school-age children to brush and floss as a preventive measure against tooth decay, leading to retention of teeth in adulthood. Over the past 40 years, dentists have shifted an emphasis of their practice from extractions and fillings to prevention of cavities. Promoting awareness in Kindergarten and grade school about brushing and flossing has prevented generations of needing false teeth, leading to better adult health through healthier gums and dentition. This paradigm-shift away from costly dental procedures such as fillings, root canals, and extractions has also paved the way for advances in oral hygiene that are linked to reduced mortality from seemingly unrelated conditions. Since the 1960s and '70s, research stemming from preventive dentistry has led to recognition that gum disease ("gingivitis") may be linked to health conditions like heart disease, blood sugar issues, as well as possibly premature and low birth weight infants. The knowledge that gum disease, chronic plaque, and abscesses lead to systemic illness in other areas of the body, such as heart disease, neonatal morbidity, and early mortality transformed the practice of dentistry from a practice of barbers in the 19th century to a highly sophisticated field of medicine today.

This creative strategy shifted dental practice toward preventive approaches rather than pulling teeth and filling cavities – transforming dentistry into the prevention realm. By promoting brushing and flossing as ways to ward off gum disease, unsightly tooth decay and tooth fillings, dentists "connected the dots" between public health and oral hygiene. Much can be learned from this upstream approach in preventing "cavities of the brain" by promoting ND-PAE prevention in reproductive health education in schools. Given the understanding that alcohol is a neurodevelopmental teratogen (chemical cause of birth defects), it is time to translate what we have learned from the science of teratology (how birth defects happen) to widespread public health strategies. The effort to promote awareness about ND-PAE as "functional birth defects" or "cavities of the brain" will hopefully inspire policy makers, legislators, physicians, and even the alcohol industry itself to shift the focus of prevention efforts upstream where the salmon are spawning. These primary prevention strategies include contraceptive use for sexually active reproductive age alcohol consumers and pregnancy planning for alcohol consumers in lieu of simply labeling alcohol with warnings for pregnant women, who may not even be aware of the pregnancy when their lifestyle behaviors are affecting their offspring.

Healthy babies are miracles in light of ubiquitous environmental chemicals that many women have a hard time avoiding for 9 months (e.g., antibiotics and growth hormone in meat and dairy products, environmental estrogens in produce and drinking water, toxins in household cleaning supplies, bi-products of plastics, lead in paint, and mercury in certain fish). With women waiting longer to have their children and significant numbers of unplanned pregnancies, it is all the more reason to become healthy before pregnancy. Agencies such as the March of Dimes promotes use of prenatal vitamins that contain micronutrients found to be neuroprotective (e.g., preventing neural tube defects, such as *spina bifida* and anencephaly). It is also known that other minerals such as zinc and iron protect developing brain cells and can further prevent anemia in the mothers. Regular exercise improves cardiovascular health and stamina for the pregnancy and post-partum period and has been shown to protect against postpartum depression and anxiety, sleep issues, and stress in parents of newborns. Other preconception and prenatal health advisories advocate for well-balanced meals with fresh fruits and vegetables. These improve metabolism and reduce the tendency to put on more weight than recommended during the pregnancy to prevent diabetes and other metabolic problems.

In today's technologically advanced, competitive world, individuals with ND-PAE migrate to the underclass of society, having difficulty sustaining employment in the working class or working poor. Unintentionally, by exposing our children to Huxley's neurotoxic solvent, silently from within the womb, we are creating a steady population destined to be stuck in poverty and reliant on public assistance – victims of society's ineptitude. Without proper diagnoses, children with ND-PAE fall through the cracks in our broken foster care and mental health system - misunderstood as children with willful behaviors, and labeled as oppositional, defiant, conduct disordered, and antisocial. As adolescents and adults, they ensure a perpetual influx of clients in a commercialized system of residential treatment programs, detention centers, jails, and maximum security prisons. I am hopeful *The Silent Epidemic* sheds light on treatment for these individuals that can change their prognosis and enable them to live happy, productive, meaningful lives.

In 1989 – 16 years after Drs. Ken Jones and David Smith had published their first paper and 8 years after the initial U.S. Surgeon General's warning about alcohol use in pregnancy – for the alcohol industry to label their product with a watered-down warning in 1989. Most educated consumers would admit the labeling is vague, difficult to read, and inadequate to warn. Unlike orange juice that cannot say "fresh squeezed" if it was produced from concentrate, alcohol is not regulated by the Food and Drug Administration and is therefore not held to the same standards as other commodities for human consumption. Clearly, the alcohol industry is not meeting its duty to warn the consumer against harm associated with alcohol use prior to pregnancy recognition.[118]

---

[118] Rich SD and Riley LJ. Neurodevelopmental Disorder Associated with Prenatal Alcohol Exposure: Consumer Protection and the Industry's Duty to Warn. Chapter 3 in *Fetal Alcohol Spectrum Disorders in Adults: Ethical and Legal Perspectives - An overview on FASD for professionals.* Nelson M and Trussler M (Eds.). Int. Library Ethics, Law Volume Number:63. Springer Publications, 2015.

Global Warning: The Missing Link between the Population Crisis and Poverty

In recent years, philanthropist billionaires have begun to study economic disparity and social issues, such as the effects of overpopulation on poverty in developing countries. In 2009, Microsoft co-founder Bill Gates convened a group of billionaires nicknamed the Good Club to join forces in social responsibility. The headlines in *The Sunday Times* read, "America's richest people meet to discuss ways of tackling a 'disastrous' environmental, social and industrial threat." The idea was simple: educate the most influential people in the world about a social cause and they begin to shift the emphasis of their individual companies and collective charitable contributions toward social revolution. Using their collective conscious, intellect, and financial resources, they are able to leverage more social change in positive ways than any government on the planet. John Harlow described how philanthropists ranging from David Rockefeller Jr, Warren Buffett and George Soros, to Michael Bloomberg, Ted Turner and Oprah Winfrey met through the leadership of the Gates Foundation to discuss ways of using their wealth to develop programs to slow population growth and improve health and education.[119]

While global overpopulation can be solved by better access to contraceptives, in the U.S., adequate contraception can also prevent ND-PAE. A recent initiative in Colorado led to free access to the intrauterine device (IUD), a relatively benign implantable method for pregnancy prevention. An anonymous philanthropist offered to pay the $600 fee for any woman to have the I.U.D. inserted, which allows up to 10 years of pregnancy-free sexual activity. From 2009-13, Colorado birth rates went down by 40% and the abortion rate dropped 35% due to fewer unplanned pregnancies, according to the Colorado Department of Public Health and Environment. State officials attributed the free access to IUDs as the primary reason for the decline in both birth rates and purposeful terminations.[120]

---

[119] Harlow J. Billionaire club in bid to curb overpopulation. *The Sunday Times*, May 24, 2009.
[120] Rubin BM. Sexually-active teens should get IUDs: doctors group. *Chicago Tribune. December 2, 2014.*

During the time the program was in place, teen and unintended pregnancy rates declined by 40%. The unnamed benefactor was unable to make his name public due to negative outcries from various special interest groups claiming that the initiative took away the reproductive rights of certain individuals. It would seem however that the women who requested to have the I.U.D.'s were doing so voluntarily and without coercion; therefore, no one segment of the population was being targeted with population control.

I learned about the Colorado program from an incredibly bright, passionate banker, Cindy - an enlightened young woman who described the disparity in teen pregnancy rates between privileged and socially disenfranchised communities. She recalled that the mothers in her affluent private high school didn't hesitate to put their teenage daughters on birth control. She said that there were no teen births in her high school, although many of her friends were sexually active, in contrast to teens living in the southeast, DC community of Anacostia whose problems with teen pregnancy stem from a desire to emancipate from their families of origin. In an online news article for Elevation DC, Christina Sturdivant outlines the complexity of the problem. She makes the point that higher teen birth rates in wards 7 and 8, which may be considered "underclass" by some socioeconomic standards, may be due to neglect by city government and socioeconomic deprivation.

> *While the teen birth rate in the District has declined 65 percent between 1991 and 2010, numbers have remained relatively unchanged over the past several years in wards 7 & 8. The impact of this crisis, however, stretches into the wallets of residents all across the city. Teen mothers often rely on public health care (Medicaid and CHIP), while their children have an increased risk of participation in child welfare and are more likely to become incarcerated—all public services by funded by taxpaying dollars. Between 1991 and 2010 there have been 24,637 births to mothers 19 and younger in Washington, D.C., costing taxpayers a whopping*

> *$1.1 billion over that period, according to The National Campaign to Prevent Teen and Unplanned Pregnancy.* [121]

Understanding that some pregnancies that may appear to be unplanned, such as teen pregnancies, are not always unintentional or mistimed. For many of the young women getting pregnant in poverty-stricken areas, they may suffer from neurodevelopmental disabilities (i.e., ND-PAE and/or learning disabilities, resulting in school failure) and believe that the best way out of the "ghetto" is to become pregnant and qualify for public assistance. The DC Campaign to Prevent Teen Pregnancy partnered with the Perry Undem Research and Communication agency to conduct a 2013 study, concluding that teens "east of the river" respond to environmental by engaging in high risk activities, such as unprotected sex, alcohol use, and other delinquent acts.

> *"...often intentionally engage in risky behavior as a response to their environment. In communities where liquor stores and drug addicts decorate street corners, teen pregnancy can be seen as a way out. ... Young people see that if I have a child, it could be a way for me to get a voucher, my own housing and supplemental money,' says Melva Williams, program director for Sasha Bruce Youthwork's Teen Outreach Program (TOP).* [122]

In my opinion, ND-PAE is the one issue that civil libertarians, republicans and democrats can share the same perspective. Our society should set limits about one's civil liberties when it comes to preventing ND-PAE. In the interest of public health, mandatory vaccinations to prevent communicable disease outweigh an individual's rights. Similarly, contraception should be mandatory and disseminated by the alcohol industry for sexually active alcohol consumers. My nonprofit, 7th Generation Foundation, has distributed condoms in an informational card at a bar and on college campuses encouraging sexually active alcohol consumers to contracept. The Better Safe than Sorry Project

---

[121] Sturdivant C. Why lowering DC's teen pregnancy rate matters to everyone. *Elevation DC, Special* Edition: Teen Pregnancy in DC. Tuesday, June 3, 2014.

[122] Sturdivant C. Why lowering DC's teen pregnancy rate matters to everyone. *Elevation DC, Special* Edition: Teen Pregnancy in DC. Tuesday, June 3, 2014.

promotes the idea that alcohol and unprotected sex don't mix. On one hand, we have civil liberties of the right to procreation, the right to choose one's lifestyle (i.e., using alcohol and other drugs), and the prolife movement. What about the rights of the unborn to a healthy brain? Healthy Brains for Children, the nonprofit founded by Jody Allen Crowe, author of *The Fatal Link*, has distributed pregnancy test kits in women's restrooms in bars in Minnesota and Alaska in order to provide a moment of contemplation for reproductive age alcohol consumers who are able to answer, "Might I be pregnant?" before drinking that bottle of beer or mixed drink.

I was a professional 20-something when I first learned about this condition. I hope that sharing my journey of enlightenment will lead to a more progressive approach to this tragically preventable epidemic.

Chapter 9

THE WAY OPENS –
ENLIGHTENMENT OF A SOCIAL DRINKER

Over 22 years ago, <u>The Broken Cord</u> provided an epiphany that prenatal alcohol exposure causes brain damage in children – even before a woman may know she is pregnant. This 'light bulb moment' sparked a fire to promote awareness and treatment of ND-PAE. - SDR

In the early 1990's as a 20-something, I found my calling for prevention and treatment of the condition now known as Neurodevelopmental Disorder associated with Prenatal Alcohol Exposure. This journey began as I was working full time in clinical neuroscience research for Burroughs Wellcome Company (BW Co.) – the maverick pharmaceutical company in Research Triangle Park, North Carolina known for bringing the first AIDS drug to the market – AZT or Zidovudine, also known as Retrovir. The "orphan drug" had shot up the FDA's approval pipeline rapidly in order to treat more patients with advanced disease. To say the least, it was an exciting and stimulating time to be a young research assistant in such an innovative company – just a year after BW Co. researchers, Drs. Gertrude Elion and George Hitchings, won the Nobel Prize in Physiology and Medicine for their discoveries leading to development of AZT. Over the span of a total of roughly 4-½ years with the company, I saw drug development from laboratory science, through the process of preclinical trials, to developing and monitoring clinical trials for Food and Drug Administration approval.

Cocooned in that unorthodox, quasi-academic, think-tank environment, my path unfolded.

The first in my family to attend college, I charted a course beyond my father's expectation for me to go to vocational school to become a secretary. Navigating upstream to a career in science, I eventually graduated *magna cum laude* from North Carolina State University with a degree in microbiology, landing a job in drug development in Research Triangle Park with Burroughs Wellcome Company. After two years in preclinical research, I left bench science to work on clinical trials. For the next two years, I enjoyed traveling to major cities, experiencing the food, culture and nightlife while monitoring pharmaceutical studies. In hind sight, the combination of unique personal perspectives – a working class upbringing, my love of science, a "social drinker's" mentality, and a maverick spirit that I could make a difference – prepared me for my awakening.

On one of my weekly trips to New York's Cornell Medical Center, I met a young researcher who opened my eyes to an appalling public health catastrophe - now understood as Neurodevelopmental Disorder associated with Prenatal Alcohol Exposure (ND-PAE). He described how this non-contagious yet preventable epidemic had destroyed the culture and diminished the intellectual capacity of indigenous communities worldwide – and had silently spread to middle class American neighborhoods. Both appalled and pathetically ignorant, I felt ambivalent listening as he described his research in pregnant rats dosed with varying amounts of alcohol at intervals during pregnancy - mimicking social drinking versus binge episodes and chronic alcoholism. As a microbiology major, I could appreciate the differences between the nerve endings in exposed rats and those of normal, healthy unexposed "control" animals. Nerve cells taken from rats born to mothers with any degree of exposure showed damage to the neural end plate, or nerve ending next to the muscle cells. Even at minimal doses, the nerve endings of the prenatal alcohol exposed pups were grossly affected – shriveled and disfigured much like gnarly fingers afflicted with rheumatoid arthritis. In contrast, the unexposed nerve cells had more delicate, intricate finger-like projections to the muscle cell. His research provided evidence for prenatal alcohol-induced nerve damage associated with sensory and fine motor issues related to sensory integration disorder.

As we discussed his findings, I was hardly moved by the results; after all, as a non-pregnant social drinker, what relevance did this have to me? Unknowingly, by sharing the implications of his research with me, that bright young researcher was forever to change my life. He handed me his copy of a book about Fetal Alcohol Syndrome (FAS), *The Broken Cord* by Michael Dorris. Half-heartedly, I agreed to look at it and return it the next week. On the plane ride back to North Carolina and in my free time the next few days, I didn't put the book down, reading it cover to cover without a break – in airport lines, grocery stores, and on my lunch break at work. I even read in the car at stoplights – the early 90's version of texting while driving. Closing the book with tears in my eyes, I understood that my life had forever changed. The way had opened for my life's purpose.

Having been raised by my paternal grandmother, who only had a 6th grade education, I had always felt that we are here for a very important purpose. She always told me – "There's nothing you can't do if you put your mind to it" and "Keep on with your education – it's the one thing no one can ever take away from you." Realizing I had found my calling, I knew I needed more education to make a difference about this problem. Still traveling and working for BW Co., I began a Master's degree in health policy and administration at the University of North Carolina (UNC) School of Public Health. There, my research focused on *The link between Fetal Alcohol Effect (FAE) and Juvenile Crime – Implications for Health and Public Policy*. Children and adolescents that do not have all the facial features of Fetal Alcohol Syndrome are less likely to be identified and receive services for their cognitive and neurodevelopmental challenges – leading to school failure, social problems, and delinquency.

Networking with North Carolina agency directors and policy makers led me to a researcher at UNC's Bowles Center for Alcohol Studies – Dr. Kathleen Sulik. Her elegant experiments in pregnant mice since 1981 demonstrated that the effects of prenatal alcohol exposure occur before most women suspect they are pregnant. Having been runner up for Miss Minnesota in 1968, Dr. Sulik is beautiful, poised, educated, and articulate – a true Alpha in Huxley's world. A modern day Renaissance woman, she is an artist as well – painting life like images of horses, people, and animals. As a young teratologist (a scientist who studies birth defects), she saw photographs of children with FAS at a Research Society on Alcoholism (RSA) conference and was inspired to better understand the mechanisms leading to the complex array of symptoms.

To that end, Dr. Sulik designed a series of experiments in pregnant mice, capturing her results in breathtaking detail using microscopic images. She had created all of the features of the full blown Fetal Alcohol Syndrome (FAS) with the equivalent of 4-5 servings of alcohol exquisitely timed to coincide with the third week after human conception.

> *When two small doses of ethanol were administered to pregnant mice during the gastrulation stage of embryogenesis, the embryos developed craniofacial malformations closely resembling those seen in the human fetal alcohol syndrome. Striking histological changes appeared in the developing brain (neuroectoderm) within 24 hours of exposure. Decreased development of the neural plate and its derivatives apparently accounts for the craniofacial malformations. The critical exposure period is equivalent to the third week in human pregnancy.*[123]

When she presented at the next Research Society on Alcoholism conference, hardly anyone believed her results. She later confided to me that most of the men in the room scoffed saying – "No woman could drink so much to raise their blood alcohol level so high." My own audiences over the years have tended to be criminologists, attorneys, and psychiatrists who counter – "Those guys haven't partied with the right women!" This little known and misunderstood fact astonished me – that major birth defects and brain damage due to alcohol can occur as early as the first three weeks of development. Of course I knew that

---

[123] Sulik KK, Johnston MC, Webb MA. *Science*. Fetal alcohol syndrome: embryogenesis in a mouse model. 1981 Nov 20;214(4523):936-8.

alcohol could cause damage to babies during pregnancy. But it was the epiphany after reading *The Broken Cord* and meeting Dr. Sulik at the University of North Carolina that fueled my passion to spread the word about this preventable tragedy. I was infuriated that my *magna cum laude* degree in microbiology from NC State University did not teach me the precarious extent of the problem–that a beverage so widely consumed as alcohol was harming babies well before most women know they are pregnant.

In my mind, having worked down the hall from Nobel Laureates, Gertrude Elion and George Hitchings – I feel strongly that Dr. Sulik should win the Nobel Prize in Medicine and Physiology for her work (in conjunction with Drs. Kenneth Lyons Jones, David Smith, and Paul Lemoine). Their independent research that alcohol can have harmful effects on offspring may actually have been a re-discovery of information known in Biblical times, during the gin epidemic in England in the mid-1700s, and in the late 1800s to early 1900's Temperance Movement leading up to prohibition (discussed in a later chapter, *Historical Perspectives*).

Part of the problem in many Native American communities, as well as other socioeconomically suppressed communities (i.e., inner city populations, educationally disadvantaged) is that alcoholism, binge drinking, and alcohol abuse are both endemic and epidemic problems. Together, with unintentional, unplanned, or mistimed pregnancies, alcohol abuse can lead to unfortunate outcomes for infants. A young woman may use alcohol before knowing she is pregnant, thereby unintentionally exposing her baby to alcohol, either because she is not using contraception or the pregnancy is otherwise unplanned. For many young women, they may have been told by their mothers, grandmothers or aunts that "I drank during my pregnancies and everything turned out okay." When people in a community become aware that alcohol can cause a problem for pregnancy, they can intervene in the lives of their family members who may be using.

When I learned about the devastation that prenatal alcohol exposure has caused indigenous communities, I became angry that very little seemed to be done to prevent it. Over the first few months working on my Master of Public Health in health policy and administration the summer of 1993, I began to see a connection between characteristics of children with prenatal alcohol exposure and those written about in delinquency literature. During that time, I frequently visited the offices of North Carolina policy makers and state agency directors while on my lunch hour. One of them, a director with the Administrative Office of the Courts, told me, "You're going to find the kids [with ND-PAE] you're looking for in detention centers and training schools." What made me so angry was that he knew about this yet nothing seemed to be done to stop the madness. I later gathered data from the Division of Youth Services and compiled a graph comparing adjudication rates by ethnicity with the population prevalence of what is now considered ND-PAE (Figure 9). A startling relationship emerged from the analysis: a similar percentage of children are incarcerated in each ethnic group as the percentage with effects of prenatal alcohol exposure.

Figure 9. NC Training School Population by Ethnicity and ND-PAE

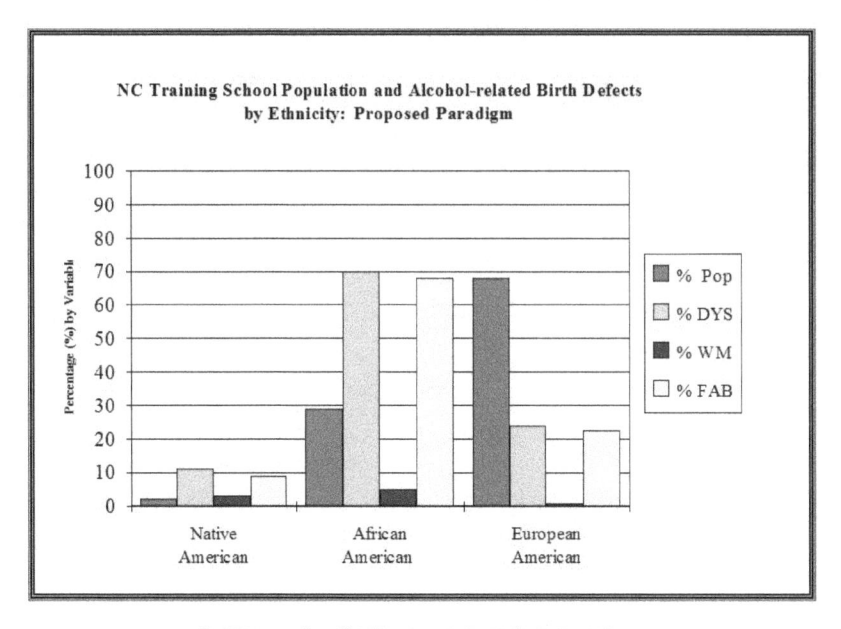

© 1994; updated 2016; Susan D. Rich, MD, MPH

While visiting the Director of Juvenile Services, I was introduced to Dr. Charles Dean, a prestigious professor of criminal justice who was heavily involved in computer mapping of high crime areas. He became perhaps my closest ally and proponent of my theory that prenatal alcohol exposure was linked to violent and persistent antisocial behavior for affected individuals who witness or experience abuse. He often said that my explanation of the link between juvenile crime and prenatal alcohol exposure accounted for 40% of the variance in his computer mapping data. Eventually, we presented together at the annual American Criminology Society and American Criminal Justice Society conferences in 1996 just before I started medical school.

Just after leaving Burroughs Wellcome to attend UNC full time, I stopped for a coffee and a biscuit on the way to work as interim Tribal Health Director for the Tuscarora Tribe of NC. In that local diner, I picked up a local newspaper with an article about the high profile murder of James Jordan, basketball player Michael Jordan's father, on the same rural road the summer before. One of the young co-defendants depicted in the article, Larry Demery, had facial features and behaviors strikingly similar to other individuals with ND-PAE. This became the first of many death penalty cases I have assisted with before and since medical school. After contacting Hugh Rogers, the young man's attorney, I was able to assist in Demery's referral to the Bowman Gray School of Medicine's pregnancy exposure team. There, he was diagnosed by a team of developmental specialists with Fetal Alcohol Effect[124] – the term used at the time to describe the neuropsychiatric and neurodevelopmental issues without the associated facial features and/or growth problems. Although evidence was presented during the sentencing phase of his trial about neurodevelopmental sequelae resulting from his mother's abuse of alcohol throughout her pregnancy, he eventually pled guilty and was given a life sentence in lieu of execution. I have detailed other cases in future chapters, choosing to share the relevant history and secondary issues related to ND-PAE without providing identifying information or gruesome details about the crimes.

---

[124] Fetal Alcohol Effects was the term used at that time to describe the outcomes in children exposed to alcohol during pregnancy who had neurodevelopmental issues in absence of either facial features or growth retardation.

A rural Native American pediatrician once used the term "Funny Looking Kid," or "FLK," to describe "dysmorphic" (abnormal) facial features of children with Fetal Alcohol Syndrome. In a compassionate way, he explained that the crass term kept physicians from understanding that prenatal alcohol exposure, rather than genetic (hereditary) factors, caused the constellation of features described by Drs. Smith and Jones in 1973. The offensive, unconscionable phrase reminded me of the off-putting language of medical paternalism in Samuel Shem's 1978 *House of God*. The revulsion learning about the way in which the most educated of our society – physicians – viewed children who could not help their disability set fire to my passion, first ignited by *The Broken Cord*. Disgust and contempt fueled the rocket that launched me to medical school– retuning to the University of North Carolina in order to bring this issue to the public's conscious. Continuing my quest to raise awareness about this preventable epidemic during medical school, I took a course in broadcast journalism and completed a summer internship at Good Morning America then produced a documentary, *Dispelling Myths about Alcohol Related Birth Defects*,[125] funded by the Centers for Disease Control and Prevention with Dr. Kathleen Sulik.

---

[125] http://bettersafethansorryproject.com/2014/01/29/dispelling-myths-about-alcohol-related-birth-defects/

© 1997, updated 2016; Susan D. Rich, MD, MPH

With a grant from the American Medical Association, as a government relations scholar, in 1998 I worked as a summer intern with the National Organization on Fetal Alcohol Syndrome – the first and only American nonprofit dedicated to raising awareness and advocating for programs to prevent prenatal alcohol exposure. Later that year, I was inducted to the NOFAS board of trustees, a position that I held through medical school, residency, fellowship, and the first few years of private practice. While NOFAS has done much to raise awareness about the harm of using alcohol during pregnancy, much is needed to increase the numbers of childbearing age individuals about the risk of alcohol use and unprotected sex.

© Susan D. Rich, MD, MPH, with permission of NOFAS and pictured individuals.
To the left: Susan Carlson, former first lady of Minnesota and a Family Court Judge (ret.).
Middle: Susan D. Rich, MD, MPH
To the right: Michele Colucci, attorney, businesswoman, and film producer, California.

Since psychiatrists see both women with alcohol and substance use problems as well as individuals with ND-PAE, I chose residency in psychiatry and fellowship training in child/adolescent psychiatry. While at Georgetown University and Children's National Medical Center, I served as Member in Training Trustee of the American Psychiatric Association, co-authoring an action paper urging the APA Assembly to consider including ND-PAE (i.e., dysmorphic and non-dysmorphic Fetal Alcohol Spectrum Disorder) in DSM-IV TR and future editions. I also worked under the mentorship of Dr. Paula Lockhart of the Kennedy Krieger Institute's Fetal Alcohol Spectrum Disorder program and went on to design a similar clinic at CNMC for evaluation and treatment of children prenatally exposed to alcohol from adoptive families, foster care, and the inner city of Washington, D.C. For the past several years of private practice, I have focused my efforts on diagnosis and treatment of this preventable disorder, receiving referrals through the National Organization on Fetal Alcohol Syndrome (NOFAS), other agencies, and word of mouth from parents of patients. These clinical experiences have shaped my perspective of how to help individuals with ND-PAE using a neurodevelopmental model rather than standard psychiatric care.

I am not alone in my interest to help these children. In addition to countless researchers and scientists who have dedicated their careers to unraveling the neurobiology of prenatal alcohol exposure, government funded clinicians across the country have studied and worked with children, adolescents and adults with ND-PAE to better understand how to diagnose and treat their condition. Recently, the Substance Abuse and Mental Health Services Administration (SAMHSA) published a compendium of this research in a book entitled, *"Treatment Improvement Protocol 58: Addressing Fetal Alcohol Spectrum Disorders."*[126] This publication provides a prevention and intervention guide for behavioral health and program administrators to develop and implement services specific to the needs of individuals with ND-PAE. The issues presented in the comprehensive manual include addressing unique treatment needs of individuals with prenatal alcohol exposure as well as helping assist women in treatment to avoid alcohol if pregnant or may become pregnant. In order for us to solve the problem of exposures prior to pregnancy recognition, we must begin to help all childbearing age women (and men) recognize their responsibility to prevent unplanned pregnancies while using alcohol.

Child psychiatrists, pediatricians, psychologists, social workers, occupational therapists, and speech pathologists are among the professionals who care for children and adolescents with neurodevelopmental disorders. Those with significant disruptive behaviors, mood dysregulation and social communication issues often are seen by child psychiatrists for medication evaluations. As a whole, the field of child and adolescent psychiatry is woefully understaffed and underfunded, with the field being in the category of "physician manpower shortage area" since the 1970's. Even if every child and teen who needed treatment were to seek services, those services would only be available in certain cities and locations in the country. There are certain states with very few board certified child psychiatrists practicing, and of those, long waiting lists for appointments. Needless to say, if the children with ND-PAE had a psychiatrist who understood

---

[126] *A Treatment Improvement Protocol 58: Addressing Fetal Alcohol Spectrum Disorders.* Substance Abuse and Mental Health Services Administration. Health and Human Services Publication No. 13-4803; 2014.

their condition, they would be in much better shape than those who fall through the cracks in the system and are left with a variety of misdiagnosed labels that neither help their prognosis nor treat their underlying neurodevelopmental condition.

With fewer than 8,500 child psychiatrists in the United States and over 75 million children and adolescents (1 in 5) needing services for a range of psychiatric and mental health conditions, a severe manpower shortage has existed since the 1970s.[127] The long years of training – four years of adult psychiatry residency plus two additional years of child and adolescent psychiatry – postpone repayment of hundreds of thousands of dollars in loans accumulated during a total of 8 years of college and medical school. In my case, I was responsible for my own educational expenses, coming from a family that was unable to support my higher education. After a two-year stint at the United States Merchant Marine Academy, I was fortunate to attend undergraduate, graduate, and medical school at state schools that are supplemented by appropriations from the North Carolina legislature. An honors program scholarship and Pell grants during college lowered my requirement for loans at North Carolina State University. As a North Carolina resident, I was also able to utilize a very special loan program through a nonprofit organization – College Foundation, Inc. that offers subsidized and unsubsidized loans to residents of the state. I also worked in a research lab at the National Institute of Environmental Health Sciences in Research Triangle Park to help offset my living and educational expenses.

My job at Burroughs Wellcome Company helped me pay off my undergraduate loans during the first four and a half years out of college. Through public health school, I worked part time on health services research projects and writing grants for nonprofits. During medical school, I was fortunate to receive a UNC Board of Governors scholarship, as well as a number of grant programs and scholarships to support my continued public health advocacy. These programs supplemented by loans and part time work, I incurred less debt than many other medical

---

[127] Thomas CR and Holzer CE. The Continuing Shortage of Child and Adolescent Psychiatrists. *Journal of the American Academy of Child and Adolescent Psychiatry*, 45:9, September 2006.

students putting themselves through school, owing around $70,000 from graduate and medical school.

Many students are not so fortunate and end up with as much debt as a home mortgage by the end of medical school. One young child psychiatrist fellow I met while on Capitol Hill for Childhood Mental Health Awareness Day and his child psychiatrist wife owe a total of $450K between them. Like them, loan repayment is a hardship for most young child psychiatrists starting out straight from fellowship training. Currently, the additional two years of training precludes child psychiatrists from qualifying for the National Health Service Corp (NHSC) loan repayment program, only available within two years after their residency – a time in which child psychiatrists are completing fellowship training. The additional expertise paradoxically makes us ineligible for loan forgiveness. With a public health background, I might have joined the NHSC and worked for Indian Health Service in order to access the loan repayment program had it been an option.

Lack of assistance for loan repayment means that a majority of child psychiatrists practice in urban and suburban areas using a fee-for-service (FFS) model. Although we are among the most highly educated in society, our average individual income of $183,637[128] is well below the median income for subspecialty training programs (Figure 5). The income disparity despite the same amount of loans leads to a higher distribution of child psychiatrists densely populated areas where psychiatrists are valued and reimbursement rates are high. Medicaid and insurance companies pay a fraction of the amount that FFS providers are able to make in private practice.

Manpower shortage and scarcity of child psychiatrists contribute to misdiagnosis and inappropriate treatment of ND-PAE. Child and adult psychiatrists see these kids every day of the week in their busy practice settings but often are unable to take the time to adequately screen them. Many psychiatrists are forced to see patients in 10-15 minute increments, often being unable to adequately assess their issues. Instead of having the time to appropriately assess the person's developmental

---

[128] http://www1.salary.com/Pediatric-Psychiatrist-Salary.html

and prenatal history thoroughly, they are left to focus their attention on symptoms and to medicate behaviors rather than looking for the underlying etiology of their condition.

Knowledge about ND-PAE, appropriate supports for their limitations, and building vocational programming around their strengths allows them to live beyond their diagnosis – much like living with a chronic medical condition. By giving their parents an understanding of how to support their challenges and provide a nurturing environment to develop their areas of strength, the child has a chance to overcome some of the obstacles they faced at birth and during early childhood. My approach is outlined in the appendices and will be further described in a companion clinical guidebook for mental health professionals to be published at a later date.

My belief is that we must direct our efforts upstream to help the children when they are still young enough to benefit from interventions and supports. Over the past several years in private practice in a home office setting, I have evaluated, treated, and cared for more than 100 patients with ND-PAE. Their capacity for understanding, judging risk, controlling impulses, appreciating nuances, navigating social situations, communicating effectively, managing a household, and other aspects of human life are more challenging than their typically developing counterparts in our modern, highly intellectualized and technical American society. Many of these individuals came early enough ages to change the trajectory of their mental health prognosis and life course.

Like my mentors at Georgetown, Children's National Medical Center and Johns Hopkins University taught me, I help patients and their parents understand their neurodevelopmental potential. The most luminary among child psychiatrists who have helped me understand these children, Dr. Paula Lockhart of the Kennedy Krieger Institute, taught me her strengths-based model that capitalizes on the child's interests and abilities while scaffolding areas of relative weakness.

Clinical, forensic, research, and advocacy efforts on prenatal alcohol exposure has led me to consult on several death penalty cases in North Carolina, Virginia, Pennsylvania, Tennessee and Georgia. My forensic work is limited to individuals with documented moderate to heavy maternal alcohol use during pregnancy. Their neuronal "hard wiring" predisposes them to fight or flight reactions precipitating violent behaviors, particularly when they have witnessed or experienced abuse as children. Prior to and during medical school, I assisted counsel for Robbie Locklear, a Native American young man convicted in the murder of his Caucasian step-father. The jury did not understand the neuroscience presented during the sentencing phase of the trial and recommended the death penalty. His case is detailed in Bonnie Buxton's eloquent memoir of her journey raising an adoptive daughter with ND-PAE, *Damaged Angels*.

Over the years, I have served as an expert in several capital murder cases in North Carolina, Virginia, Georgia, Pennsylvania, and Tennessee involving defendants with significant neurodevelopmental deficits resulting from prenatal alcohol exposure. Although the Maryland death penalty was repealed a few years ago and I was never consulted on any of the five men who were on capital murder charges in the state, I have evaluated a variety of young people in detention centers and residential treatment programs, with histories of moderate to heavy prenatal alcohol exposure. Some of these unfortunate individuals have well-intentioned adoptive parents or biological mothers who are now in recovery from substance abuse yet have little knowledge about the impact of prenatal alcohol use on their child's brain development. Somehow, no one in the system of care treating their child felt compelled to share the important information with the parents. These experiences reinforced the findings from my Master's thesis that undiagnosed effects of prenatal alcohol exposure put children who witness or experience abuse at risk of violent and persistent juvenile delinquency behaviors. Sadly, not enough effort is spent educating parents and caregivers about the need to provide cocooning, protect from harsh parenting strategies, and assist youngsters in developing a moral compass as well as vocational skills at an early age.

Having just returned from the American Psychiatric Association (APA) in Toronto, Canada, I have hope that our profession is finally at a place where change is on the horizon. There, I chaired a symposium on ND-PAE as a silent epidemic of preventable mental illness. A more senior colleague and distinguished life fellow commented during the question-answer session that it was one of the most illuminating talks of the entire conference yet held on the last day with few attendees still present to hear it. Notwithstanding his frustration, I believe the way will open for future presentations on a larger scale, similar to sessions I've presented with colleagues at the American Academy of Child and Adolescent Psychiatry. Such a paradigm shift will surely help enlighten mental health professionals about prenatal alcohol exposure as a source of acquired brain damage that can be treated, though more easily prevented.

There is hope in a tipping point presented in the next chapter, *Doctor's Responsibility to Prevent ND-PAE*, to prevent neurodevelopmental disorder associated with prenatal alcohol exposure – comparable to mainstream practice for prevention of prenatal exposure to other drugs and medications.

Part II

Shifting Social Paradigms

Chapter 10

Doctors' Responsibility in ND-PAE Prevention

Alcohol has been used for centuries by the medical community as a pharmaceutical drug, both for "treatment" and "prevention" of illness. Over the past 80 years since the repeal of prohibition, physicians have encouraged pregnant women to use alcohol to ease the pain in their backs, to help them relax and sleep better at night, and to help their "milk drop" (i.e., stimulate lactation). Intravenous alcohol drips were an ineffective though frequent practice in the obstetrics community throughout the past century to prevent preterm labor.[129] Such historical precedence has engendered a lack of physician acceptance that alcohol is a teratogen.

Reflecting on an excerpt from the modern Hippocratic Oath, rewritten in contemporary terms and context,[130] we find a paradigm shift toward prevention, recognition of socioeconomic impact of illness on families, and

HIPPOCRATIC OATH, MODERN VERSION

I swear to fulfill, to the best of my ability and judgment, this covenant:

I will respect the hard-won scientific gains of those physicians in whose steps I walk, and gladly share such knowledge as is mine with those who are to follow.

---

[129] Abel EL. A critical evaluation of the obstetric use of alcohol in preterm labor. *Drug and Alcohol Dependency.* 1981 Jul;7(4):367-78.

[130] Written in 1964 by Louis Lasagna, Academic Dean of the School of Medicine at Tufts University, and used in many medical schools today.

I will apply, for the benefit of the sick, all measures which are required, avoiding those twin traps of overtreatment and therapeutic nihilism.

I will remember that there is art to medicine as well as science, and that warmth, sympathy, and understanding may outweigh the surgeon's knife or the chemist's drug.

I will not be ashamed to say "I know not," nor will I fail to call in my colleagues *when the skills of another are needed for a patient's recovery.*

I will respect the privacy of my patients, for their problems are not disclosed to me that the world may know. Most especially must I tread with care in matters of life and death. If it is given me to save a life, all thanks. But it may also be within my power to take a life; this awesome responsibility must be faced with great humbleness and awareness of my own frailty. Above all, I must not play at God.

*I will remember that I do not treat a fever chart, a cancerous growth, but a sick human being, whose **illness may affect the person's family and economic stability. My responsibility includes these related problems,** if I am to care adequately for the sick.*

I will prevent disease whenever I can, *for prevention is preferable to cure.*

I will remember that I remain a member of society, with special obligations to all my fellow human beings, *those sound of mind and body as well as the infirm.*

If I do not violate this oath, may I enjoy life and art, respected while I live and remembered with affection thereafter. May I always act so as to preserve the finest traditions of my calling and may I long experience the joy of healing those who seek my help.

— *Louis Lasagna, MD; Tufts University, 1964.*

In keeping with this and the former version of the Hippocratic Oath, we physicians are bound by a moral and ethical obligation to help inoculate individuals who may be using alcohol during reproductive years and their families against inadvertent prenatal alcohol exposure by improving access to the primary prevention information. As we are among the most educated members of our society, we have a duty and responsibility to admit what we do not know instead of telling patients that we know of no harm in alcohol use during reproductive years or during pregnancy. We are typically the warriors on the front lines of contagious outbreaks of any illness on the planet – recognizing the signs and symptoms and alerting public health practitioners of the pending epidemic. We are also among those infantry officers who help contain the outbreak and improve access to care for the infirmed. Instead, our profession is falling short of its primary responsibility to the greater good of society, shirking our obligation to prevent disease.

Still, more than 40 years after identification of Fetal Alcohol Syndrome in the US, medical myths about alcohol continue to be perpetuated by doctors – based less on hard science and more on individual clinical practice. While alcohol is no longer being prescribed as a drug, many physicians are complacent about the impact of alcohol on the developing fetus or about their duty to warn childbearing age women about the potential harm of PAE. Some physicians not only fail to inform women about the potential for alcohol-related brain damage prior to pregnancy recognition, there are those still telling women that "a little alcohol is okay" during pregnancy, "just don't overdo it." I recall several women I encountered during medical school while working as an intern with the National Organization on Fetal Alcohol Syndrome who were told by their fertility specialists that they could have "a few drinks every now and then." The problem with such generalities is that there is a large range in our society for what is considered socially acceptable drinking. One drink may be a sip of wine to one woman and a Long Island Iced Tea containing six shots to another. Cutting back on alcohol may mean going from a fifth of vodka a day to a pint or several mixed drinks per day to a few per week.

Obstetricians tend to see pregnant women well into their first trimester – around the 8th to 10th week of pregnancy. By that time, much damage may have already occurred to a fetus. This prenatal care practice is based on the statistic that 30% of conceptions end in spontaneous abortion, mostly in the first trimester; therefore, if a woman presents "too early" after conception, she might have had a "miscarriage" by the next office visit. Preconception care visits are not a routine part of gynecological care and often are not covered by insurance companies. While preconception screening, referral and education could benefit all childbearing age women, lack of insurance reimbursement for these primary prevention visits often prohibits gynecologists from providing such care.

Pediatricians are well positioned to screen children for ND-PAE during routine office visits, yet they are often unaware of obstetric records, even for women who are high risk alcohol and substance users. Depending on the socioeconomic status of a woman, the child's provider may not even ask the woman whether she drank alcohol – particularly prior to pregnancy recognition. In most cases, children with ND-PAE are not identified by physicians unless they have obvious facial features of FAS. Often, geneticists erroneously believe it is a death sentence more than a helpful diagnosis – as though the patients have a chronic, end stage illness preventing them from any productive and meaningful life. A number of young adults with ND-PAE I have evaluated in prisons and detention centers had earlier diagnoses of Fetal Alcohol Syndrome but their parents had been told that there was nothing that could be done so there would be no need to tell the individual about the diagnosis. The unfortunate truth is that ND-PAE predisposes individuals to hidden disabilities that make it difficult to navigate everyday life in a family much less the complexities of an increasingly complex society.

Some women drink daily in keeping with the American Heart Association guidelines – which used to be two drinks per day, but has been modified to one drink daily for women or two for men. The guidelines now clearly state that women at risk of pregnancy should abstain from alcohol altogether. However, that message is only minimally referenced on the website and there has been so much media promotion and advertising about the benefits of alcohol use that many women have become confused about the risks, particularly in the face of an unintended or mistimed pregnancy. In short, because of mixed messages from health care professionals and the media, women are not getting clear information to make an informed choice about the impact of their lifestyle behaviors on their children's future cognitive potential and mental health. Inadequate warnings about the dangers of unintended pregnancy and alcohol use has led to an epidemic number of children with some degree of ND-PAE.

My colleague, Laura Riley – a bright young attorney in Los Angeles and I have written articles and discussed the heated debate over whether to blame, shame or incarcerate a woman who uses alcohol during pregnancy. We both believe that the focus of responsibility should be shifted to the medical community and government agencies that share a duty to warn consumers with the alcohol industry. The following edited excerpt was omitted from our book chapter entitled, "Neurodevelopmental Disorder associated with Prenatal Alcohol Exposure: Consumer Protection and the Industry's Duty to Warn," in *Fetal Alcohol Spectrum Disorders: Ethical and Legal Perspectives.*[131] Reviewers commented that a more narrowly directed focus of the chapter with tightly referenced citations would allow the article to have more impact than a broader perspective.

Individual medical professionals, societies and organizations have a duty to encourage their members to comply with the U.S. Surgeon General's updated advisory in 2005, in keeping with the ethical and moral duty to deliver sound medical advice to their patients. Recently, the American Academy of Pediatrics and the American College of Obstetrics and Gynecology have joined forces with the Centers for Disease Control and Prevention to promote awareness to pediatricians about the dangers of PAE (AAP, online publication). The American

---

[131] Rich SD and Riley LJ, Neurodevelopmental Disorder Associated with Prenatal Alcohol Exposure: Consumer Protection and the Industry's Duty to Warn. Chapter 3 in *Fetal Alcohol Spectrum Disorders in Adults: Ethical and Legal Perspectives - An overview on FASD for professionals.* Nelson M and Trussler M (Eds.). Int. Library Ethics, Law Volume Number:63. Springer Publications, 2015.

College of Obstetrics and Gynecology recognized their responsibility in prevention of ND-PAE, publishing the following position statement:

> *For pregnant women and those at risk of pregnancy, it is important for the obstetrician–gynecologist to give compelling and clear advice to avoid alcohol use, provide assistance for achieving abstinence, or provide effective contraception to women who require help. (ACOG, 2011).*

It is clear from these guidelines that the medical societies concur with the U.S. Surgeon General's advisories that no amount of alcohol used during pregnancy is safe, even prior to pregnancy recognition.[132] Medical professionals are gatekeepers of information to help inform those in their care of potential harm due to lifestyle behaviors, not only informing them about risks of treatment for illnesses. This process of informed decision making includes providing enough information to a patient to help the patient make a reasonable decision about their medical treatments, health outcomes, and choices (in this case, lifestyle choice to use alcohol in the face of contrary medical information). According to the American Medical Association's Code of Medical Ethics "informed consent" [133]

> *The physician's obligation is to present the medical facts accurately to the patient ... and to make recommendations for management in accordance with good medical practice ... Social policy does not accept the paternalistic view that the physician may remain silent because divulgence might prompt the patient to forego needed therapy.* (Issued March 1981).

When applied to the issue of preconception information regarding alcohol teratogenicity, medical providers are implicitly obligated according to the Code of Medical Ethics to inform all childbearing age patients about the dangers of alcohol use during early stages and throughout pregnancy. This obligation includes providing information about the known and suspected risks of social drinking on reproductive health outcomes, including the

---

[132] Surgeon General's Advisory on Alcohol Use in Pregnancy. *MMWR Weekly.* March 11, 2005; 54(09);229.
[133] American Medical Association's Code of Medical Ethics, 1996-97; p. 120.

potential risk of inadvertently harming a fetus. Regarding patient information, the Code of Medical Ethics (AMA, 1996-97; p. 125) states:

> *It is a fundamental ethical requirement that a physician should at all times deal honestly and openly with patients. Patients have a right to know their past and present medical status and to be free of any mistaken beliefs concerning their condition ... Only through full disclosure is a patient able to make informed decisions regarding future medical care ... Concern regarding legal liability which might result following truthful disclosure should not affect the physician's honesty with a patient. (Issued March 1981).*

In keeping with the Hippocratic Oath to "do no harm," health care professionals should also avoid statements to pregnant women such as, "a little alcohol is okay." Sound medical advice is that any amount of alcohol can be harmful at any stage in pregnancy. Based on anecdotal information gathered from 1-800 pregnancy exposure hotlines and surveys of obstetricians, many doctors continue to give mixed messages and minimize alcohol's effect on fetal development. Recently, a study from Great Britain reported that two drinks of alcohol a week during pregnancy is actually "helpful for offspring." Given such mixed messages in the media, it is important that government agencies and physicians be clear about the messages they are giving to pregnant women and adequately inform them of the potential damage, given the overwhelming evidence to the contrary.

As conduits of health information, medical providers have a duty to warn alcohol-consuming patients about alcohol's teratogenic effects (i.e., about the potential risk of ND-PAE), just as they educate about a wide range of lifestyle behaviors and decisions. [There are realistic limits to the extent of this obligation – e.g., in the case of an alcoholic patient.] What then about health care professionals who may not be informed themselves about the adverse effects of alcohol on pregnancy, particularly with regard to dose-response mechanisms as well as early, binge, and frequent exposures? According the Code of Medical Ethics, continuing medical education is vital to the practice of medicine:

> *Physicians should strive to further their medical education throughout their careers, for only by participating in continuing medical education (CME) can they continue to serve patients to the best of their abilities and live up to professional standards of excellence.* (AMA, p. 136).

Thus, the old adage that "ignorance is innocence" does not hold up to this credence. No matter when a physician graduated from medical school (whether before or after PAE was known to be teratogenic), it is not a physician's privilege to plead ignorance in light of insurmountable knowledge about the effects of alcohol on reproductive health outcomes. In a general way, the Code of Medical Ethics addresses physicians who might avoid disclosing to women risks of alcohol use potentially damaging a fetus in the event that they know the woman may have been drinking before knowledge of the pregnancy. When a new obstetrical patient asks a physician, *"Could my baby have been hurt by the drinks I had at a New Year's party last week?"* according to the Code of Medical Ethics, a physician is bound to explain all the risks of alcohol use despite the guilt that the woman may experience, in order that she be "free of any mistaken beliefs."

In the event that a physician or his/her colleague has told a pregnant woman that "a little alcohol is okay," has failed to warn her of the potential consequences of use, and has failed to discuss her drinking patterns with her, from an ethical view point, s/he should not withhold additional information from her. She should be informed about the possible outcomes of her drinking on her baby's health status and the probability that the early binge exposures may have caused school age mental impairment or minimal brain damage.

Mixed messages in the media come in forms that sometimes seem legitimate health advice. The American Heart Association (AHA) promotes health benefits of light to moderate alcohol use, suggesting that the risks are outweighed by potential health benefits (AHA, online publication). Yet even the AHA's website lacks information about ND-PAE, stating "Pregnant women shouldn't drink alcohol in any form. It can harm the baby seriously, including causing birth defects." They clearly are promoting an outdated message. Another related area of intense debate is the use of alcohol during lactation. Mixed information in the media lead patients to seek clarification from their physicians about what is safe before and during pregnancy, as well as lactation. One seemingly credible, yet profoundly inaccurate online source states:[134]

- *Wait at least two hours after you finish a drink before nursing your baby to give your body a chance to clear the alcohol.* [If alcohol's half-life is 12 hours, how can one be certain that the amount of alcohol you consumed 2 hours before is already cleared from your system? The physiology doesn't add up! Also, what if the woman's one drink is a cocktail with several shots or a Long Island Iced Tea, containing 6 shots?]
- *Your blood alcohol level (and the level of alcohol in your milk) is generally highest 30 to 90 minutes after you have a drink, although that time – and the length of time it takes the alcohol to leave your body – varies from person to person.* [This seems to be their disclaimer to avoid risk of law suit from a "bad baby case."]
- *You can time your drink so that your baby won't be nursing for a few hours afterward by having it right after a feeding, for example, or during one of your baby's longer stretches of sleep.* [You can guarantee that the baby will sleep better if you wait 2 hours after consuming alcohol – by then, there certainly still will be appreciable alcohol in the system.]
- *Or you can pump and store your milk before having a drink, then feed your baby expressed milk from a bottle. (Pumping after you drink won't clear alcohol from your system any faster – it will still take at least two hours.)*
- *Another option is to feed your baby formula in the hours following your alcohol consumption.* [Now they're talking common sense. If you can't avoid alcohol entirely while breastfeeding, maybe formula will be a better option. The benefits of breast milk are outweighed by the potential brain damage caused by nursing after alcohol consumption.]
- *To ward off dehydration, down a glass of water in addition to the alcoholic drink. It's also a good idea to eat beforehand or*

---

[134] The Baby Center, online article. http://www.babycenter.com/0_alcohol-and-nursing-moms_3547.bc

when you're having your drink. This helps lower the amount of alcohol in your blood and your milk.

Such information may confuse even the most educated women, leaving the impression that it is okay to drink alcohol during breast feeding. On the other hand, because every person's metabolism and alcohol clearance rate is different and every baby's susceptibility to the damaging effects of alcohol on the brain is unique, it is nearly impossible to say that alcohol use within 24 hours of breastfeeding is safe. Physicians need to educate themselves and their patients about the dangers in promoting use of alcohol during reproductive years without a caveat to patients that they should be using contraception if drinking, and to avoid alcohol while planning pregnancy, pregnant, or breast feeding.

Medical professionals should screen all childbearing age populations for alcohol use, educate them about the potential for unintentional alcohol-related birth defects and cognitive deficits resulting from unplanned pregnancies, and encourage them to use reliable contraception if sexually active and using alcohol. Women should be asked about their drinking and drug use habits prior to, not just during pregnancy and at delivery. This ethical duty to screen reproductive age women for alcohol use levels is similar to standard of care for assessing HIV risk factors or risk of teen pregnancy. It is also standard of care for women prescribed known teratogenic medications (neuroleptics, retinoic acid products, mood stabilizers, anticonvulsants, chemotherapy, etc.) to use two reliable forms of contraception and to plan their pregnancies carefully to prevent fetal exposure to these drugs.

The aim of shifting the paradigm to primary prevention is not to expose physicians to unwarranted "bad baby" malpractice claims. Instead, we hope to inspire physicians to take the time to encourage sexually active childbearing age alcohol consumers to use contraception. Professional medical associations and state licensing agencies have at least an ethical obligation to advocate that childbearing age women be screened and referred to treatment based on their risk of PAE. Where indicated, physicians should screen and refer patients with alcohol problems to community resources, especially those planning a pregnancy, who are unable to stop on their own (as a method of preconceptional reproductive counseling). This spirit of healthier futures – with alcohol-free pregnancies from the entire preconception (before pregnancy) period through the end of lactation (breast feeding).

Chapter 11

NATURE VERSUS NURTURE ... OR BOTH!

> *Genes alone do not explain the complex patterns of inheritance observed for psychopathology: both genes and environment contribute ... Historically there has been a long-standing interest in the impact of rearing environment on mental health. ... now it is more widely accepted that many potential environmental risks, such as life events and parenting, are heavily influenced by genetically influenced characteristics of parent and offspring,[135] who shape, select, and evoke environmental circumstances.[136]*

The debate over whether a child's neurodevelopment is more impacted by hard-wiring at birth (nature) or affected by environment (nurture) is not as simple as black and white: is it nature OR nurture? It really is both.[137] Genetics play a large role in not only hair and eye color or metabolism and body size, but also in a person's intelligence, stress response, immune function, emotional reactivity, and social relatedness. Yet the outcome of hard-wiring the nervous system is not purely genetic either. The electrical wiring and hormonal processes are further compromised by exposure to certain "neurodevelopmental teratogens" – chemicals that influence brain function. Alcohol is one of these substances – probably the most ubiquitous and easily accessible

---

[135] Plomin R. Genetics and Experience: The Interplay between Nature and Nurture. Sage series on Individual Differences and Development, Vol 6. Thousand Oaks, California, Sage Publications, 1994.

[136] Thapar A. Parents and Genes and their Effects on Alcohol, Drugs, and Crime in Triparental Families. Editorial. *American Journal of Psychiatry.* 172:6, June 2015.

[137] Collins WA, Maccoby EE, Steinberg L, Hetherington EM, Bornstein MH. Contemporary research on parenting: The case for nature and nurture. *American Psychologist,* Vol 55(2), Feb 2000, 218-232.

yet least understood by society. Prenatal alcohol-induced faulty wiring of neurons in the brain and nervous system can change the natural, genetically-programmed interactions between infant and caregiver during infancy and early childhood. Early institutionalization, a chaotic/violent home life and neglectful/abusive parents are extrinsic (environmental) factors that further influence development.

Figure 10. The Multi-hit Model of Neurodevelopmental Disorders:[138] The interplay between genetic factors, prenatal alcohol exposure, and "postnatal" experiences during childhood and adolescence contribute to one's "neurodevelopmental phenotype."

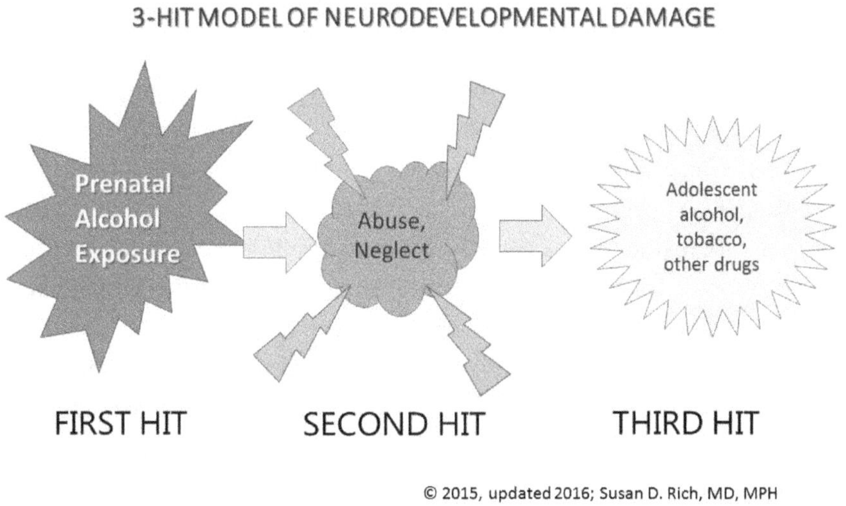

3-HIT MODEL OF NEURODEVELOPMENTAL DAMAGE

Prenatal Alcohol Exposure → Abuse, Neglect → Adolescent alcohol, tobacco, other drugs

FIRST HIT　　SECOND HIT　　THIRD HIT

© 2015, updated 2016; Susan D. Rich, MD, MPH

The diagram in Figure 10 illuminates the complex interactions between one's underlying genetic background (gene's and epigenetic factors), prenatal alcohol exposure, and life experiences. As depicted, genetic factors (i.e., maternal and fetal risk and protective genes) along with other prenatal and postnatal influences contribute to the resulting "phenotype," or neurodevelopmental profile of an individual. In my experience working with children who have the diagnosis of ND-PAE, deficits inherent in the condition itself puts an individual at increased

---

[138] Adapted from Picci G and Scherf KS. A Two-Hit Model of Autism: Adolescence as the Second Hit. *Clinical Psychological Science*. May 2015; vol. 3 no. 3:349-371.

risk of negative outcomes. However, in absence of a diagnosis of ND-PAE, their risk is much higher. Without an adequate diagnosis, they lack the environmental supports to assist with their continued development and may remain at risk in their environment due to secondary brain damage from abuse and neglect. This second hit of neurodevelopmental damage happens as a result of witnessing or experiencing abuse, neglect, or loss of mothering – putting the child/adolescent at even greater risk than their typically developing peers with the same adversity.

Their neurodevelopmental phenotype is further shaped by their interaction with the environment during adolescence when rewiring of brain circuits occurs. Often, parents, caregivers, and teachers inadvertently over react to their outbursts, perpetuating the fight or flight reactions in a feedback loop of stressàreactionàmaladpative responseàfurther stressàreactionàand so on. In this highly stressed, highly reactive state, experimentation with alcohol, tobacco, and other drugs of abuse occurs, leading to the "third hit" to the developing brain (adolescence-onset substance abuse), further limiting the individual's capacity for resilience and health. The second and/or third hit then worsens the prognosis by further diminishing the capacity of the biologically-challenged system. Hence, conditions such as mood disorders (i.e., bipolar disorder), thought disorders (i.e., schizophrenia), and other forms of mental illness manifest in a higher frequency of patients with "multiple hits" to their neuronal wiring and neurotransmitter systems.

To provide context for this multi-hit phenomenon, underlying genetic causes of functional and physical birth defects (microdeletions, Triple X syndrome, Trisomy 21, Fragile X) may lead to worse outcomes for embryos with prenatal alcohol exposure. For example, a child who might have had genetic "loading" for autism because both parents had Asperger's-like traits will have even more severe effects because of prenatal alcohol exposure. Likewise, a woman with a genetically slow metabolism of alcohol due to differences in her cytochrome P450 enzyme systems carrying an embryo that may have genetic susceptibility to the effects of alcohol would have a worse outcome, possibly even with lower amounts of alcohol. Higher blood alcohol levels are associated with worse outcomes than lower levels, although in a genetically vulnerable mother-fetal pair, lower amounts of alcohol could be harmful. One child in my practice has Triple X Syndrome and Fetal Alcohol Syndrome. She was adopted at birth from another state by loving parents who later learned about the diagnoses from a geneticist who referred her to me for treatment. Because of the underlying genetic disorder (Triple X), she is tall and gangly in stature and had a tendency toward some cognitive deficits. The prenatal alcohol exposure further compromised her "phenotype" or "expression of the genes" by causing a more dysmorphic, or odd-appearing, face and more significant brain damage than might have been seen only with the genetic condition. At the age of 8 years old, she functions at the level of a 4-year-old academically, much younger socially, and emotionally has inconsolable "terrible 2's temper tantrums" that are exhausting for her parents to predict, prevent, preempt, or prepare for.

The amount of alcohol consumed will affect each mother-fetus pair differently. A woman may interpret her physician's advice that "a little is okay, just don't overdo it" to mean "mild to moderate use of alcohol is okay." This slippery slope mentality can easily lead to risky drinking, with "a little" being interpreted as cutting back on her current consumption of a bottle of wine with dinner to one or two glasses. A woman who binge drinks on an empty stomach obviously will have a higher blood alcohol concentration (BAC) than a woman who has a sip or two of alcohol with dinner. Conversely, a "sip or two" can be interpreted broadly if allowed by a health care provider. For some women and/or their fetuses, their physiology is such that the alcohol may be metabolized much slower. Certain ethnic groups are at higher risk due to a genetic variant in alcohol dehydrogenase enzyme which dramatically reduces the ability of the body to eliminate alcohol. In a newly pregnant woman (e.g., during the embryonic period in the first 8 weeks post conception), those differences in metabolic rate may magnify the effects of the alcohol. The amniotic sac ("womb") holds the alcohol longer while it is processed through the fetal circulation (lacking adequate enzymes to digest the alcohol), leaving the baby exposed to greater concentrations than in the mother's system. Women who are simultaneously using other substances or have been exposed to environmental chemicals may have worse outcomes for their pregnancies due to a combination and/

or synergistic effect on the child. Certain chemicals reduce protective factors while others may cause their own teratogenic effects.

The ugly truth is that babies born to women who consume moderate to heavy amounts of alcohol regularly are born smelling like a bar the night after the super bowl. The amniotic fluid becomes a fermenting station, a pickling plant of sorts. In my view, it is a miracle that many of the babies survive the alcoholic womb at all. For women who also smoke cigarettes (which constricts the uterine vessels, cutting off the oxygen supply and other nutrients from the blood), their babies will be at even higher risk for effects of prenatal alcohol exposure. Video images taken by sonogram of babies exposed to maternal smoking show the babies covering their eyes with their hands, as though to avoid the smoke. Women who use other medications or illegal/recreational drugs, are exposed to environmental chemicals, have poor nutrition, are stressed by domestic violence or other life events, work long hours, or experience other difficulties have worse outcomes than those who are relatively healthier prior to and during pregnancy.

In research animals, binge amounts of alcohol can cause limb deformities similar to those caused by thalidomide, though other more lethal birth defects caused by the early exposure leaves few survivors. In humans, the combination of genetic risk, alcohol, and other environmental exposures may increase this unfortunate outcome. One boy in my practice adopted from Latvia has severe phocomelia (lack of hands and forearms), low-set ears and moderate intellectual disability. His Russian birth parents, immigrants to Latvia, drank heavily before, during and after the pregnancy then abandoned him in the hospital after birth due to his disfiguring birth defects. He was eventually placed in an orphanage for many years before he was adopted at age 4. A keen attachment therapist noticed features of FAS in his face as well as neurodevelopmental challenges and recommended that I see the child. I sent a photo of him to Dr. Kathleen Sulik and Dr. Ken Jones asking whether they believed prenatal alcohol exposure could have played a role in his limb deformities. Dr. Jones did not believe that prenatal alcohol contributed to the child's condition and felt it was more likely a genetic condition. After a couple of weeks, Dr. Sulik replied with a photograph of a mouse fetus indicating "alcohol induces severe distal upper limb reduction defects." Her note read,

> In light of the recent inquiry re: the patient with upper limb reduction defects, I thought you both might be interested in the attached figure which was made from some mouse fetuses that we 'delivered' yesterday for an ongoing study. We have known for some time that in mice, if the timing of exposure is just right (GD 9.25; approximately equivalent to days 26-28 post-fertilization in humans), alcohol can cause significant loss of the postaxial digits [fingers] and forearm. Typically, the right limb is more frequently and more severely affected than the left. The lower limb remains apparently unaffected. You might also note in the figure that the right eye is smaller than the left, a finding indicative of brain damage. The face is not that of typical FAS as the timing of insult was too late to cause that pattern.
>
> I agree that the child in question may very well have a (solely) genetic condition. Or, it could be a genetic abnormality that is exacerbated by alcohol exposure. Based on animal studies, an influence of alcohol on this type of defect is certainly within the realm of possibility.

Another colleague, Dr. Ronald Federici, an internationally renowned neuropsychologist and adoptions expert, mentioned that the limb deformities could also be related to prenatal exposure to any number of heavy metals in the ground water in Latvia. He reminded me that many Soviet-era nuclear weapons were improperly disposed of and leak their remains into the soil and water table in Eastern European countries. There residents are unknowingly exposed to hazardous chemicals and inadvertently expose their offspring to the toxins. He also reminded me about the range of other environmental toxins less regulated and more widely accessible to human contact in Eastern Europe and Russia. Dr. Federici has travelled extensively to most of those areas, bringing back several children he adopted from some of the most depraved conditions one could imagine an orphanage to be. He also established the Bucharest Early Intervention Project[139] – a revolutionary foster care program in

---

[139] http://www.bucharestearlyinterventionproject.org/

Romania which has proven successful in improving the outcomes for children placed into families prior to age 2.

Herein lies the "nature AND nurture" phenomenon. The interaction between an individual's biological and physiological characteristics from birth (nature) and their environmental conditions (nurture) leads to their constitutional "phenotype" (personality, character, responsiveness, aggressive tendencies, etc.). In this view, nature comes somewhat from genetics and partly from prenatal influences, whereas nurture provides the elements influencing the child from birth through adulthood. After birth, brain development (nature) is positively or negatively influenced by environmental (nurture) factors. For example, positive influences on brain development during the first three years of life include loving, nurturing supportive caregivers; early enrichment with music, art, books, socialization; and calm, safe, predictable surroundings. Since many children live their first three years of life in multiple foster care homes or orphanages, the "Zero to three" years represent lost opportunity for brain development and in many cases, worsening of their brain function. Recent studies show that child abuse negatively effects brain development similar to physical head trauma.[140]

Research and common sense suggest that only a subset of children who witness or experience abuse go on to have violent or delinquent behaviors. Another caveat, fortunately, is that not all children prenatally exposed to alcohol have neurodevelopmental impairment. In turn, not all children with ND-PAE become juvenile offenders or "criminal." It is the "synergistic" effect between neurodevelopmental issues (i.e., social communication, neurocognitive, emotional regulation, sensory integration, and adaptive functioning) combined with abusive and traumatic experience that predispose an individual to delinquent behaviors. For individuals with "neurodevelopmental deficits" who have witnessed or experienced abuse during childhood, they are at higher likelihood of delinquent, antisocial, or violent behaviors. The early childhood trauma (e.g., long term institutionalization, family violence,

---

[140] Falcone T, Janigro D, Lovell R, Simon B, Brown CA, Herrera M, Myint AM, Anand A. S100B blood levels and childhood trauma in adolescent inpatients. *Journal of Psychiatric Research*. Published Online: December 24, 2014.

frequent or traumatic disruptions in their primary caregivers and/or catastrophically chaotic home environments), leave individuals with ND-PAE at greater risk of other psychiatric illness that may precipitate physical or emotional outbursts in times of stress, frustration, or anger. That is, the combination of faulty neurodevelopmental functioning ("hard wiring problems"), exacerbated by early trauma experiences leave children vulnerable or "at risk" of antisocial tendencies.

Although the full impact of the social environment is beyond the scope of this book, the interaction of the childhood experience on individuals with ND-PAE cannot be overlooked. Briefly stated, ND-PAE during infancy, toddlerhood, childhood, and adolescence predisposes an individual to worse adverse psychological outcomes from early institutionalization, parental loss, physical/emotional/sexual abuse, neglect, and other forms of trauma. The implications of ND-PAE on development, behavior, academic and adaptive functioning over the life span can be best understood in the context of the interaction of social and familial factors with an individual's neurodevelopmental deficits. Early institutionalization, neglect, abuse, and family violence may engender different presentations in this population, depending on the quality and degree of underlying neurodevelopmental impairment. For these reasons, care must be taken to tease out symptoms based on developmental versus social history in order to develop a complete, appropriate understanding of the interplay between brain-based and environmental (post-natal) origins of psychopathology.

There is strong evidence linking the stress levels of the mother, her state of "arousal" to the child's reactivity to stress. Mothers and fathers who read to their babies in the womb and minimize stress in their lives typically have babies who respond to their parents voices after birth, are more attentive when read to and are more self-regulated. So, the baby interacts with the outside world (environment) even before birth. Children with neurodevelopmental issues who have experienced or witnessed significant trauma, physical or emotional abuse, or profound institutional neglect are at risk for hyperarousal ("fight or flight") responses to environmental triggers. Animal research has shown that prenatal alcohol exposure increases the influence of mild chronic stress on the development anxiety-like disorders in adulthood.[141]

Hence, the impact of ND-PAE (see Appendix A: *What is ND-PAE?*) on human potential must be considered in the context of one's caregiving environment. A disconnect ensues between nature and nurture, leaving the infant vulnerable to difficulties with care givers, resulting in oppositional and defiant behaviors. For an individual with ND-PAE who is abused or victimized, either overtly or vicariously through media, they are even less likely to be able to empathize with the victims of their own actions. Often, even after the catastrophe, they are unable to appreciate the impact of their actions in terms of remorse, guilt, or shame. Physical trauma within the home or school (i.e., child abuse, domestic violence, bullying) or vicariously through violent videogames, "slasher" movies, and killing in television programs "conditions" to tolerate, accept, and prefer such behaviors.

In my experience working with patients, the home and school environments play a role in helping these children and adolescents develop appropriate self-regulation. It is helpful to minimize chaos, improve structure, diminish expectations of academic achievement, and shift to a vocational track or a hands-on apprenticeship model as opposed to a diploma track. The idea that every American child should graduate high school and attend college far exceeds the capacity of many of these individuals. More realistic expectations would be accomplishing activities of daily living such as personal hygiene, cleanliness, self-care, meal preparation, regulated sleep and wake cycle, laundry, and other chores. These adaptive functioning skills can be developed through vocational training into a good work ethic – showing up on time, maintaining a job, paying bills on time, balancing a checkbook, grocery shopping, meal planning) should be seen as realistic goals for many of these individuals. Developing meaningful and enduring friendships, being married to the same person for a lifetime rather than for a few

---

[141] Hellemans KG, Verma P, Yoon E, Yu W, Weinberg J. Prenatal alcohol exposure increases vulnerability to stress and anxiety-like disorders in adulthood. *Ann N Y Acad Sci.* 2008 Nov; 1144:154-75.

months or years are important aspects of life that these individuals often miss out on.

The way an infant responds to touch, taste, temperature, textures, sounds, smells and other sensory inputs to facial expressions, subtle rhythms, vibrations, transitions and nuances of the environment is affected by prenatal alcohol exposure. The resulting interplay between a child's underlying neuronal wiring and their caregiving environment becomes the foundation for development of personality, psychopathology, and/or emotional health. In infants with ND-PAE, underlying faulty neuronal wiring can cause emotional dysregulation, maladaptive attachment behaviors, and disruptions in attunement with others. These issues can be associated with disruptions in attunement and attachment behaviors all the way through childhood and adolescence.

Their fragile nervous systems need protective cocooning from birth through adolescence and into adulthood to accomplish the important rewiring that ultimately can occur. Having the right network of providers and community supports in place to augment the care and nurturing environment provided by the family will go far in helping these children thrive. We repeat the phrase in child psychiatry: "It takes a village to raise a child" as though it is a mantra for parents to recite, yet without much understanding or guidance to parents what that means. It means to seek out mentoring from healthy role models like coaches, clergy and young adults in the community. When the state and federal-run mental hospitals were deinstitutionalized from the 1960s to the 1980s, individuals were put into the communities on "chemical restraints" (i.e., high potency antipsychotics and other medications) with poor community-based programs to assist those young people from a psychiatric perspective. At the same time, there are a number of systems that are in place for management and support of these children but lack a coordination of care (i.e., a ship without a captain).

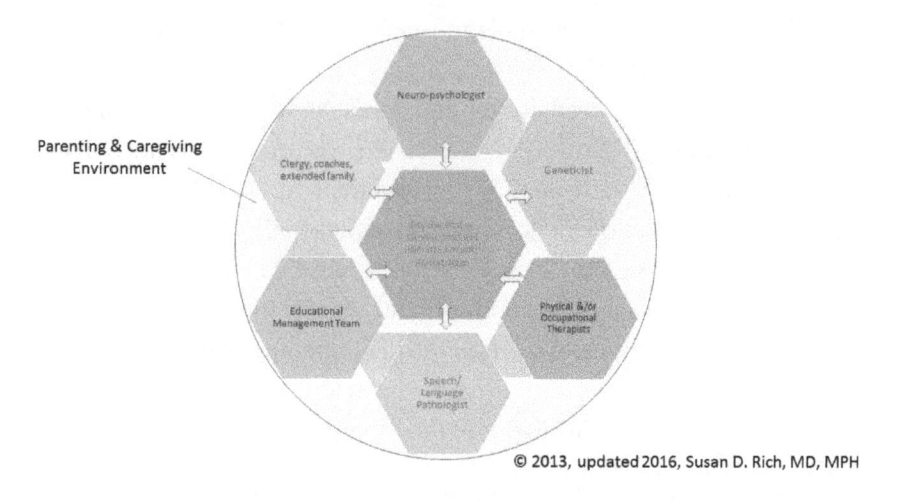

Figure 11. Multidisciplinary Approach to ND-PAE Treatment – Scaffolding the Cocoon

© 2013, updated 2016, Susan D. Rich, MD, MPH

I believe a child psychiatrist or developmental pediatrician can help manage the multiple care providers, to provide appropriate supportive therapies either through insurance sponsored programs or community-based (e.g., county run) services, and to use medication as an augmenter of environmental supports, rather than the other way around. By providing the proper cocoon to keep them safe and not using any substances during adolescence, their nervous system has a much better time of rewiring itself in ways described in this book.

There is a healing power of time without the added deficits from trauma during early childhood and additional substance abuse during adolescence as depicted previously in Chapter 10. I tell children in my practice: "You've had enough alcohol to last 10 people 10 lifetimes, my friend. All before you were born. So, if anyone ever offers you alcohol or other drugs, simply tell them – 'No, thanks. I've had enough already.'" As I have developed a relationship with them since an early age, they will listen to this mentoring approach very differently than to their parents who they may perceive as overprotective, controlling, or overbearing. If they will listen to the advice and avoid the alcohol during this period of rapid brain re-organization, they will avoid the third neurodevelopmental hit during adolescence. Unfortunately, research shows that prenatal alcohol exposure causes conditioning for alcohol and other drug abuse during adolescence and young adulthood. Left to their own devices and "free will," the child will grow into a gratification-seeking, self-medicating adolescent who experiments with then becomes addicted to alcohol and other substances of abuse. Their once loving, tolerant, and supportive parents become frustrated, helpless, and demoralized as they watch the decline of the child they put so much of their love, faith, dreams, and hope into when they adopted them. In a sense, they are traumatized by not only the physical aggression they endure but the anguish of loss of the potential that "might have been" had their child remained substance-free.

There need to be more alcohol and other drug abuse treatment centers focused on understanding and treating ND-PAE. So often, the child, adolescent and young adult is unable to navigate the system of care as it is today. Many "treatment" systems are based on one's ability to understand abstract concepts, distinguish consequences, regulate emotions, and control behaviors. From my clinical perspective, deficits in emotional regulation and mood, implicit ability to comprehend the nuances of social situations, auditory or visual information processing, functional working memory, and/or other executive functions make it difficult to develop the skills necessary to

Individuals with ND-PAE at risk for further psychopathology in the face of environmental stressors. Individuals with FASD are particularly vulnerable to environmental stressors (e.g., early parental loss, trauma, institutionalization, witnessing or experiencing abuse). Those lacking social supports, coping skills, and other resources that promote resiliency are often faced with school failure, disruptions in foster care placements, delinquent behaviors, and adjudication. Often, teasing out the neurodevelopmental (prenatal neuronal hard wiring) pathology from the trauma-related (postnatal neuronal programming) can help clinicians design a unique treatment plan targeted to the underlying wiring issues rather than "shooting from the hip" at the symptoms.

If identified and treated early, with proper supports and structure as well as a stable, nurturing environments and realistic expectations, these children have the potential to live happy, productive lives – albeit with significant limitations. Parenting approaches presented in the appendices offer my approach to strategies for helping these children. First, adoptive and foster parents must ask the right questions and be given accurate answers about the prenatal histories of children they are caring for – unlike so many well-intentioned families I have met who knew little about the maternal/prenatal records and even less about how the child may have been affected. Adoption and child welfare agencies must take the time to adequately train their staff and parents who have the means and heart to provide the right supportive, nurturing environment for neurodevelopmentally compromised children.

Communities must be willing to use resources efficiently and strategically to provide realistic, targeted services to young women who may have been in foster care without significant family support to aid her parenting and/or have ND-PAE, which may limit her social and cognitive capacity to effectively parent her children. It will take creative, well-funded community-centered strategies to overcome the cycle of poverty, alcohol and substance dependency, and multiple pregnancies plaguing the foster care population. Through the same innovation and philanthropy that birthed the child welfare movement, we will relieve the social ambivalence and moral dilemma that has perpetuated the problem.

The next chapter, *Augmenters to Foster Care,* recommends strategic, well-planned vocational and housing programs for individuals with ND-PAE and their mothers in order to break the cycle leading to poverty, multiple children with alcohol/other drug exposures, homelessness, and reliance on public assistance.

Chapter 12

AUGMENTERS TO FOSTER CARE

> *Hospitals have become revolving doors for "acute stabilization," but when brain-damaged individuals break the law, they tend to be confined immediately and receive few if any services. Deinstitutionalization of the mental health system and lack of community services has led to a default system of care for this vulnerable population (i.e., detention and youth facilities, jails, and prisons). – Natalie Novick Brown and Susan D. Rich*[142]

Recently, a disturbing conversation with a young disability law attorney led to my understanding just how broken our system is today. We were discussing my clinical findings about his client, a 20-year-old African American young man ("Ed") – adopted from foster care at age 4, who is applying for disability services. Another of his 20-year-old clients ("Devon"), "aged out" of foster care before being adopted. The Department of Developmental Disabilities had denied Devon services, leaving another branch of the government to deal with his needs: criminal justice. I shivered as he explained Devon would be left to the revolving door of jails and prisons to deal with his adaptive functioning deficits. While both he and Ed have ND-PAE with moderate intellectual disability, Devon lacks adoptive parents have resources

---

[142] Brown NN and Rich SD. A Neurodevelopmental Paradigm for Fetal Alcohol Spectrum Disorder. *The Judge's Page* published online by the Court Appointed Special Advocates and National Council for Juvenile and Family Court Judges; Winter 2014.

and the education to fight for housing, vocational supports, and other services to assist in his transitioning to adulthood.

Far too many of these young people who age out of foster care before being adopted are warehoused today in residential treatment, detention centers, and training schools as adolescents – and fall off the proverbial iceberg into the icy waters of death row. The current system is not working. It is a human rights tragedy that we have allowed a prison system to evolve into a corporate economy, earning revenue from the misfortunate cognitively disadvantaged, illiterate individuals who are unable to navigate the complexities of our society.

My hope is that our society is on the verge of a radical paradigm shift – akin to the enlightenment in child welfare at the beginning of the 20th century. A glimpse into this history provides insight into how society continues to fail our most vulnerable children. In that era, children were considered chattel – a parent's property much like ownership of a slave. If a child died in the care of a parent, most cases did not lead to investigation of malintent. A benevolent movement in children's rights spawned the fields of child protection and child psychiatry. The child protection movement began in the late 19th century when (ironically) the founder of the Society for Prevention of Cruelty to Animals, Henry Bergh, and his attorney, Elbridge Gerry, helped remove an orphan girl from an abusive foster home in 1875 and later created the New York Society for the Prevention of Cruelty to Children (NYSPCC). Their advocacy in child welfare thus spawned a number of other nongovernmental child protection societies across the country. John E.B. Myers, who has written a number of books on the topic,[143] provided the following evocative history in a 2008 article in *Family Law Quarterly*:

> Organized child protection emerged from the rescue in 1874 of nine year-old Mary Ellen Wilson, who lived with her guardians in one of New York City's worst tenements, Hell's Kitchen. Mary Ellen was routinely beaten and neglected. A religious missionary to the poor named Etta Wheeler learned

---

[143] Myers JEB. *Child Protection in America: Past, Present and Future*, 2006 and *A History of Child Protection in America*, 2004.

> *of the child's plight and determined to rescue her. Wheeler consulted the police, but they declined to investigate. Next, Wheeler sought assistance from child helping charities, but they lacked authority to intervene in the family. At that time, of course, there was no such thing as child protective services, and the juvenile court did not come into existence for a quarter century. Eventually, Wheeler sought advice from Henry Bergh, the influential founder of the American Society for the Prevention of Cruelty to Animals. Bergh asked his lawyer, Elbridge Gerry, to find a legal mechanism to rescue the child. Gerry employed a variant of the writ of habeas corpus to remove Mary Ellen from her guardians.* [144]

> *The history of child protection in America is divisible into three eras. The first era extends from colonial times to 1875 and may be referred to as the era before organized child protection ... intervention to protect children was sporadic, but intervention occurred. Children were not protected on the scale they are today, but adults were aware of maltreatment and tried to help [as was the case of Mary Ellen Wilson and Etta Wheeler]. The second era spans 1875 to 1962 and witnessed the creation and growth of organized child protection through nongovernmental child protection societies. The year 1962 marks the beginning of the third or modern era: the era of government-sponsored child protective services.* [145]

Around the same time, child psychiatry was birthed out of concerns about the issue of juvenile delinquency. Dr. John E. Schowalter detailed the early history of Child Psychiatry in a 2003 *Psychiatric Times* article:

> *Most historians of child psychiatry date its beginning in this country to 1899, when Illinois established the nation's first juvenile court in Chicago. This occurrence set forth the following sequence of events. A group of influential, socially concerned women on the board of directors of Jane Adam's Hull House was shocked by juvenile delinquency. They wanted to understand its origin, prevention and treatment. These women were approximately 90 years ahead of the*

---

[144] Myers JEB. A Short History of Child Protection in America. *Family Law Quarterly*, Volume 42, Number 3, Fall 2008; pp. 451.
[145] Myers JEB. Fall 2008; pp. 449-463.

Centers for Disease Control and Prevention's decision to accept violence as a public health problem. In 1909, these foresighted women created the Juvenile Psychopathic Institute and hired a neurologist, William Healy, M.D., to be its first director. Although a neurologist interested in studying the delinquents' brain functioning and IQ, the perspective of the settlement house's board of directors made sure that attention also was paid to the delinquents' social factors, attitudes and motivations. To accomplish these broad evaluations and treatment strategies, Healy formed teams composed of a neuropsychiatrist, a psychologist and a social worker. This approach became the template used by most child guidance clinics for most of the 20th century. Child psychiatry's roots became implanted in the community, rather than in medical schools, and colleagues were more likely to be teachers, judges, social workers and social scientists, rather than physicians.[146]

A Solution or Part of the Problem?

In my clinical experience as well as while working in rural, NC, a majority of children with ND-PAE tend to be one of several children born into the same family – sometimes with multiple fathers, often knowing their younger siblings fathers better than their own. One of the difficulties in this endemic condition that has reached epidemic proportions is that the mothers themselves often have prenatally-induced brain damage. Remember Stephanie from the chapter *Predator versus Prey* (p. 120)? In addition to having early sexually provocative experiences, Stephanie spent much of her time eating food out of the trashcans and with an excoriating diaper rash from wearing soiled diapers all day. Both she and her baby sister, Bella, were profoundly developmentally delayed and malnourished and – partly from nature (prenatal alcohol exposure) and partly nurture (lack thereof).

The foster family who took the girls had won awards for their care of other children. However, the Jenkins family who later adopted the girls learned that their foster parents had locked Stephanie in a closet

---

[146] Schowalter JE. A History of Child and Adolescent Psychiatry in the United States. *Psychiatric Times*, September 01, 2003.

during her meltdowns and tantrums because they were at a loss for what to do to keep her and others in the house safe. No one in the system explained what neurodevelopmentally challenged children need in parenting styles or provided appropriate training to ensure they were experienced enough to deal with the emotional dysregulation. The other problem with the system is that there are no "stop gaps" for births to women who abuse alcohol in pregnancy. Stephanie and Bella have a total of six younger half siblings, all born to their mother over the years since they were adopted. Each time she lost custody of one, she went on to have another child. Hearing that she herself grew up in foster care, I asked if the Jenkins had a picture of Stella. Now at the age of 32, she is overweight yet still shows characteristic stigmata of prenatal alcohol exposure – small, wide set eyes, elongated philtrum, thin upper lip, short upturned nose, flattened midface.

One might say that feminists are right to respect the reproductive health wishes of women. I agree. However, we must all acknowledge that this vicious cycle of unintended pregnancy (or sometimes intentional in order to become independent of her family of origin) and alcohol and other drug abuse must end. In her cohort of 415 individuals with ND-PAE, Dr. Anne Streissguth found that 30 women had given birth to a total of 55 children –with half the children (54%) no longer in the care of their biological mothers and 30% of the children removed by child protective services. An alarming 40% of the women had drank alcohol during their pregnancies.[147]

Supportive Parenting

Children with ND-PAE from socioeconomically suppressed backgrounds tend to come from unstable family environments, and many end up in foster care, a factor that increases their chances of entering the juvenile and criminal justice systems. The National Organization on Fetal Alcohol Syndrome reports that up to 70% of children in foster care have ND-PAE. During fellowship training at

---

[147] Grant TM, Ernst CC, Streissguth AP, and Porter J. An advocacy program for mothers with FAS/FAE. In: Streissguth AP and Kanter J (eds.) *The Challenge of Fetal Alcohol Syndrome: Overcoming Secondary Disabilities*; p 102-112. Seattle: University of Washington Press, 1997.

Children's National Medical Center in Washington, DC, a clerkship with Dr. Paula Lockhart at the Kennedy Krieger Institute in Baltimore, MD and in my private practice in Potomac, MD, I have focused on this population. An understanding of the neurodevelopmental domains discussed in the appendices can help scaffold and cocoon children, adolescents and adults with ND-PAE in order to promote healthier outcomes than found in longitudinal studies such as those conducted by Dr. Anne Streissguth in Seattle, WA.

Advances in social consciousness led to reforms in child welfare laws of the 1980s and 1990s, placing orphaned children into U.S. families through a greater commitment to adoptions and increased foster care. Children previously hidden away in institutions were returned to the community. From the 1960s to the present in the U.S., orphaned, neglected, and abused children have been adopted or placed in foster care out of an altruistic commitment to child protection. An understanding of this history is relevant to the reasons that our system is terribly broken – contributing to neurodevelopmental challenges of ND-PAE and attachment disorders that make it so difficult for children to thrive in our ever complex modern society. The following brief timeline of the laws around foster care and adoptions is paraphrased from the National Coalition for Child Protection Reform: [148]

- ➢ In 1961, Congress allowed children in foster care to receive Aid to Families with Dependent Children (AFDC) payments, making foster care cheaper for states and localities, and dramatically increasing the foster care population.
- ➢ Late 1970s: Foster care population reaches 503,000. Congress becomes concerned that too many children in poverty are being placed out of the home unnecessarily due to poorly defined "neglect."
- ➢ 1980: Adoption Assistance and Child Welfare Act of 1980, successful in that it cuts the foster care population to 243,000. Notable as the first law to: require reasonable efforts to keep

---

[148] National Coalition for Child Protection Reform. *Setting the Record Straight on Recent Child Welfare Reform: A Child Welfare Timeline;* Sept. 12, 2010.

families together; set time limits on how long children could stay in foster care (18 months); encourage permanence by either returning children to biological parents or making adoption an option; and the first to offer federal aid for subsidized adoption. Foster care remained an open-ended federal entitlement for states and local government, capping the minimal funds to prevent foster care.

- 1992: The U.S. Supreme Court rules that individuals cannot sue to have the "reasonable efforts" requirement enforced (i.e., child welfare workers being required to make reasonable efforts to return the child to the biological parents).
- 1993: The Family Preservation and Support Act (now called "Promoting Safe and Stable Families Act") is enacted and funded for $1 billion distributed over five years, covering a range of child welfare services – foster care, adoption, and after school recreation programs.
- 1997: The Adoption and Safe Families Act is enacted, favoring termination of parental rights when a child has been in foster care for 15 of the previous 22 months. Small increases in adoptions occur until 2000.
- 2000: Plateau in the number of adoptions from foster care.
- 2003: Number of children coming into care falls below where it was when ASFA became law
- 2005: A record 307,000 children placed in foster care.
- 2008: Increases in adoptions annually equals no more than 2.5% of the children in foster care on any given day. The number is outweighed by the children coming into care.
- 2010: Many foster children "age out" of the system before finding permanent homes. Child abuse declines.

In truth, the foster care system grew out of society's attempt to keep children safe from harm. However, it has become a faulty, poorly designed "lifeline" for children who have already been prenatally exposed to alcohol and other drugs and/or scarred physically and emotionally by months or years of abuse and/or neglect. The 2009-2012 Report to Congress on Child Welfare Outcomes (CWO Report) by the Children's Bureau of the U.S. Department of Health and Human Services Administration for Children and Families illuminates the characteristics of child victims:

> *In 2012, there were approximately 679,000 instances of confirmed child maltreatment. The overall national child victim rate was 9.2 child victims per 1,000 children in the population. State child victim rates varied dramatically, ranging from 1.2 child victims per 1,000 children to 19.6 child victims per 1,000 children. While the national child victim rate decreased from 9.3 child victims per 1,000 children in the population in 2009 to 9.2 in2011, there was no change between 2011 and 2012. Child victim rates in 2012 varied rather substantially across racial/ethnic groups. Black children had the highest rates of victimization at 14.2 victims per 1,000 children in that racial group's overall child population. Asian children had the lowest rates, with 1.7 victims per 1,000 Asian children in the population.*

The CWO Report further describes information about the foster care population:

> *Nationally, there were approximately 397,000 children in foster care on the last day of 2012. During that year, an estimated 252,000 children entered foster care, and 241,000 children exited foster care. Among the states, the foster care entry rate ranged from 1.3 children per 1,000 to 8.6 children per 1,000 in a state's population. Between 2002 and 2012, the number of children in care on the last day of the FY decreased by 24.2 percent, from 524,000 to 397,000. While it is currently not possible to determine the cause of the decrease in the number of children in foster care using the AFCARS database, a number of states have been making deliberate efforts to safely reduce the number of children in care through various programmatic and policy initiatives. The rates of children in foster care in 2012 varied substantially across racial/ethnic groups. American Indian/ Alaska Native children had the highest rates of children in care, with 13.0 per 1,000 children in that racial/ethnic group's overall child population. Asian children had the lowest rate, with 0.7 in care per 1,000 Asian children in the general child population. Nationally, 235,000 children exited*

foster care in 2012. Of these children, 207,000 (87 percent) were discharged to a permanent home (i.e., were discharged to reunification, adoption, or legal guardianship).[149]

From government reports compiled by the Child Welfare Information Gateway, the impact of parental substance abuse on the child welfare system can be summarized as follows:

An estimated 12 percent of children in this country live with a parent who is dependent on or abuses alcohol or other drugs. Based on data from the period 2002 to 2007, the National Survey on Drug Use and Health (NSDUH) reported that 8.3 million children under 18 years of age lived with at least one substance dependent or substance-abusing parent. Of these children, approximately 7.3 million lived with a parent who was dependent on or abused alcohol, and about 2.2 million lived with a parent who was dependent on or abused illicit drugs. While many of these children will not experience abuse or neglect, they are at increased risk for maltreatment and entering the child welfare system. NSDUH is an annual SAMHSA survey of a representative sample of the national population. It defines dependence and abuse using criteria specified in the Diagnostic and Statistical Manual of Mental Disorders (DSM), which includes symptoms such as withdrawal, tolerance, use in dangerous situations, trouble with the law, and interference in major obligations at work, school, or home over the past year. The most recent data analyzed related to children of substance abusing or dependent parents are from the 2002 to 2007 surveys.

For more than 400,000 infants each year (about 10 percent of all births), substance exposure begins prenatally. State and local surveys have documented prenatal substance use as high as 30 percent in some populations. Based on NSDUH data from 2011 and 2012, approximately 5.9 percent of pregnant women aged 15 to 44 were current illicit drug users. Younger pregnant women generally reported

---

[149] Child Welfare Outcomes 2009-2012 Report to Congress. *Safety, Permanency, Well-being.* U.S. Department of Health and Human Services Administration for Children and Families; Administration on Children, Youth and Families; Children's Bureau. (http://www.acf.hhs.gov/programs/cb/resource/cwo-09-12)

the greatest substance use, with rates approaching 18.3 percent among 15- to 17-year-olds. Among pregnant women aged 15 to 44 years old, about 8.5 percent reported current alcohol use, 2.7 percent reported binge drinking, and 0.3 percent reported heavy drinking.[150]

Grace Court: An Enlightened Alternative

Instead of waiting until babies have experienced significant prenatal brain damage and/or postnatal trauma (child abuse/neglect) to intervene, a more upstream approach would be provision of wrap around services (parenting education, vocational screening and job placement, substance abuse services) within a supportive housing community for individuals affected by alcohol and other drugs. In the ideal situation, there would be improved programs to promote mental well-being, screening for lifestyle behaviors that are risky, and referral for childbearing age individuals to receive appropriate services prior to pregnancy, not just after pregnancy recognition. In this way, individuals who have substance abuse problems, mental health conditions, or ND-PAE would receive adequate supports from society until they are able to parent their own children independently.

An enlightened, socially responsible program is one such program providing alternatives to foster care for homeless women with substance abuse problems and their dependent children. Grace Court is a remarkable 24-unit transitional housing community in Robeson County, NC – a rural, predominately tri-racial community. As Director/Developer of Special Programs for Robeson Health Care Corporation (RHCC), my position involved supervising 25 substance abuse counselors and social workers, writing grants, and developing programs for the consortium of 5 community health centers. RHCC also operates Our House, a halfway house for pregnant women– which was one of four established in the late 1980s by a grant from the Substance Abuse and Mental Health Services Administration's Center for Substance Abuse Treatment. The clients are

---

[150] The Child Welfare Information Gateway. *Parental Substance Use and the Child Welfare System* Bulletin for Professionals; October 2014. Specific publications referenced in the report are omitted for readability. https://www.childwelfare.gov/pubs/factsheets/parentalsubabuse.cfm)

referred by neighboring counties as well as the local community and typically enter at about their fourth to fifth month of pregnancy – well after much damage to their fetuses had occurred. They were able to spend the next year and a half, until their baby's first birthday, living in the house, learning parenting skills, working or attending school, and having transportation and assistance for medical appointments. Many were mandated by the courts or encouraged to be there by their families in order to have the comprehensive system of care provided to them. On or before their child's first birthday, often to the same substance abusing environment they came from., they were assisted to transition back to their communities –. Invariably, some of the women would return to Our House with another pregnancy two or three years later – having lost custody of their baby RHCC had helped them care for.

The Executive Director of RHCC – a genius public health entrepreneur – was the brains behind the community health center and the halfway house. She wanted to know why many of the women seemed to be on a vicious cycle of multiple substance exposed pregnancies. During the process of strategic planning to develop the initiative, I held focus groups with the women to get their input and suggestions; some were living at Our House and others had completed the program one or more times. When I asked why they kept returning to Our House with an additional substance exposed pregnancy, the women said essentially, "Because the government keeps taking my babies away from me." Essentially, they were returning to their same drug-infested neighborhoods, relapsing, and losing custody of their child.

After presenting what I had learned to the Executive Director, we decided to create an alternative approach to scaffold and support the biological mothers, thereby improving their existential need and desire to parent their own children. Envisioning a 24-unit transitional housing facility for women in recovery and their dependent children, she hired a housing consultant who developed an array of creative funding streams to build a beautiful gated community – 24 apartments with 3-4 bedrooms each. I wrote the grants and developed the programs for the transitional housing complex, including a $300K grant for Special Needs Homeless Populations from the U.S. Department of Housing and

Urban Development. Residents transition from Our House with their toddlers and are reunited by the courts with their children in foster care – parenting the children they had rather than perpetuating the cycle of neglect, parental loss, and trauma leading to issues with attachment.

"Grace Court" – named by the residents after the phrase *"There too go I but for the Grace of God"* – was established in the impoverished rural community at the midway point on the drug trafficking route from Florida to New York becoming the first of its kind on the East Coast. I continued working on that project and a few others in development for RHCC even after beginning medical school at the University of North Carolina. Grace Court opened around February 1998, my second year of medical school. It was a joy and privilege to work with such an amazing group of socially responsible free thinkers – building a program for women in recovery to sustain healthy lifestyle behaviors, attain an education or an improved vocational experience, and parent their children in a supportive community. Other innovative strategies to shift the paradigm toward parents have been presented in a document produced by the Child Welfare Information Gateway.[151]

One might argue that putting such carefully designed and socially conscious systems in place to help support young mothers in their parenting of their children will be costly and unrealistic. However, I often say that we built it from the ground up in 2 years – a $1.8 million transitional housing community. The supportive services the women receive are no costlier than the social service benefits they would otherwise be receiving at the County agencies. The same programs are offered on site in a "one stop shop" approach where women can walk over after work or before school for therapy, to meet with their case manager, or to have a supportive parenting session.

The cost of these housing communities are substantially less than foster care, recidivist delinquency, substance use behaviors, and jail for these innocent children – brain damaged by our social drug of choice. There are alternatives to this penguin march off the iceberg into death row due to their inability to understand consequences, being

---

[151] Available online at https://www.childwelfare.gov/pubs/factsheets/parentalsubabuse.cfm

gullible, easily led and influenced, and having maladaptive coping strategies. Part III shares perspectives for professionals and parents to screen, intervene and shift toward positive parenting of affected children, including recognition of their strengths, supporting their areas of challenge, and providing ways for them to live in society with a protective cocoon while their brain continues to develop.

Chapter 13

SOCIETAL SOLUTIONS TO THE SILENT EPIDEMIC

Ironically, we humans enjoy combining alcohol and sex, which can lead to unintended pregnancy and inadvertent PAE. Since over half of pregnancies are unplanned, it is misguided to think that it is a woman's responsibility to know the risks of drinking during her childbearing years, or indeed while she is pregnant, when health professionals and governmental agencies are providing insufficient—or in some cases incorrect—information. It is in the interests of childbearing age patients and society to prevent ND-PAE.

Scientists and the public health community have known about the intellectual disability, learning problems, low birth weight, infant death, and other life-long consequences of alcohol for several generations and have chosen a variety of inadequate measures to address it. Prohibition did not work, although largely influenced by the effect of maternal alcohol use on infant death and child mortality. Not even advisories issued by the U.S. Surgeon General have shifted the way we think about alcohol from a reproductive health context in the last 31 years – the focus has remained on women who are pregnant or planning a pregnancy. It is not enough to warn pregnant women about the use of alcohol since the point of pregnancy recognition is already too late to prevent much damage to the wiring of the brain.

To ensure the duty to warn society adequately about the harmful effects of alcohol prior to pregnancy recognition, consideration should be given to reintroducing a derivative of *S. 2047 (100th) "Alcoholic Beverage Labeling Act of 1988."* Since alcohol as a beverage is both a food and a drug, it should be regulated by the Food and Drug Administration – not one of the poorest funded, smallest government agencies (i.e., Alcohol, Tobacco and Fire Arms). The state of Alaska has taken on the epidemic of ND-PAE by placing pregnancy test kits in bars and restaurants where alcoholic beverages are sold, with the aim of raising awareness about the dangers of alcohol use for women who may be pregnant. While pregnancy tests in bars may increase pregnancy recognition, the most reliable results come from a morning urine sample. In the least, the test dispensers will have women "stop and think" about whether to have that next drink or even about their reproductive potential while at the bar. Maybe it will inspire them to use reliable contraception or to avoid alcohol if they have been having unprotected sex.

Recently, the Centers for Disease Control and Prevention issued a report in its *Morbidity and Mortality Weekly* encouraging alcohol consumers to use reliable contraceptives.[152] A number of media sources reported backlash from women who found the advisory off-putting, sexist, and alarmist,[153] even though within a week of the recommendation, the Zika virus outbreak lead to both the CDC and the Pope himself encouraging women in endemic countries to prevent pregnancy. The time has come for physicians to encourage their sexually active, childbearing age, alcohol consuming patients to use reliable contraception. Medical providers should work closely with other human service professionals to ensure that populations consuming large amounts of alcohol (alcohol abusers and alcoholics) who may be less able to abstain from alcohol use without intervention have access to substance abuse treatment (outpatient, inpatient, transitional housing, etc.). Every effort should be made to address the contraceptive health needs of these very high risk childbearing age individuals.

Another, albeit more "upstream approach," would be distribution of condoms with an informational flyer about ND-PAE for the first alcoholic drink of the evening sold to each customer. Signing "informed

---

[152] *Vital Signs*: Alcohol-Exposed Pregnancies — United States, 2011–2013. *Morbidity and Mortality Weekly*; February 5, 2016 / 65(4);91–97.

[153] http://www.npr.org/sections/health-shots/2016/02/04/465607147/women-blast-cdcs-advice-to-use-birth-control-if-drinking-alcohol

consent" to acknowledge that a person understands and is willing to accept the risks to the unborn of unintentional exposure to an early embryo would allow another point of education to young people. A tax on alcohol by the bottle or by the drink imposed by local and state liquor boards then earmark proceeds to these ND-PAE prevention initiatives would help communities stem the tide on this preventable disorder. These monies could also be used to educate and provide medical and psychiatric care for affected individuals. The educational programming, supports, parent guidance, allied health therapy, and psychiatric care for individuals affected by prenatal alcohol can be costly even for adoptive upper middle class families– the modern day "social service agency." What about birth families bearing the costs of the expenses for these children? Many are in the working class and working poor with little means to support their families' basic needs let alone expensive health care and educational costs. By the time the child reaches school age, the parent is often so frustrated by the child that they would have been abused, neglected, and feel disenfranchised from the family. More about parenting strategies can be found in Part 3, Appendix C: Parenting a Child with ND-PAE.

To that end, the following points support stronger legislation and action:

> Alcohol use even before a woman may know she is pregnant can cause physical, cognitive, emotional, psychiatric and behavioral problems for her child. While ND-PAE may occur with more than minimal prenatal alcohol exposure, the full blown Fetal Alcohol Syndrome (FAS) occurs as early as the late third to early fourth week of pregnancy with as little as a 4-5 servings over the short course of an evening.

> At least half of all US pregnancies are mistimed or unplanned; that is, the couple is not planning to become pregnant. This statistic has remained relatively consistent over the past 3 decades despite advances in family planning. Therefore, preconception initiatives are critical to reducing inadvertent exposure to alcohol early in pregnancy.

- Approximately 14 million women in the US binge drink up to 3 times per month with about 6 drinks per episode. If only 10% of these women became pregnant, the outcomes would be catastrophic.
- The number of women who drink alcohol while pregnant has consistently remained about 1 in 8, according to a 15 year-study by the Centers for Disease Control and Prevention and approximately 2% engaged in binge drinking or frequent use of alcohol.
- The results of the CDC Study further indicated that more than half of women who did not use birth control (and therefore might become pregnant) reported alcohol use and 12.4% reported binge drinking.
- Rates of ND-PAE exceeds or rivals that of a number of other non-preventable concerning neurodevelopmental disabilities (autism, cerebral palsy, Down's syndrome, and *spina bifida*).

Since it is estimated that one in twenty American children have ND-PAE, even if we were able to identify all of them, communities lack adequate neurodevelopmental services needed to address all of their complex needs. Much of the time, school systems are reluctant to continue Individualized Education Plans (IEPs) for children transitioning from the early childhood intervention programs into Kindergarten. One child in my practice who had been "kicked out" of preschool for disruptive behavior was unable to get an IEP for Kindergarten despite my urging the school administration that it could help prevent "fight or flight" reactions. Within a few weeks of school starting, the school staff had called the police on the 5-year-old who had cocooned himself under the teacher's desk and became agitated when the teacher tried to coax him out to complete his school work. In a SWAT-team style drill, the teacher cleared the room of students, called in the janitor and PE teacher who pulled the boy out from underneath the desk and held him down until the police arrived. What had been a frightening ordeal that led him to hide under the desk became a traumatic experience for him and his twin brother who watched the entire event transpire while he stood watching helplessly from the hallway.

Truly, it takes a village – every village – to plan for, birth, and nurture children in a supportive, safe, and humane environment. This will mean rethinking the idea of "safe harbors" and sanctuaries in pastoral settings – the initial concept of asylums, for these socially and emotionally fragile human beings. From a local community-based perspective, creating alcohol and substance-free housing for families of origin to raise their own children will be a giant leap toward fixing our fractured foster care system and breaking the vicious cycle of early childhood trauma. As opposed to fostering children to be raised by strangers, we need to rethink policies that foster parents by offering safe and drug free residential communities and minimizing marketing of alcohol to low income, minority, and socially disenfranchised populations. Children, adolescents and families with ND-PAE need greater access to mental health care, targeted approaches to treatment, and improved community-based resources – including educational and vocational programs, healthy social outlets, and recreational activities. Such programs are much less costly than lives lost due to unrecognized and untreated neuropsychiatric issues. Societal priorities must shift toward the mental well-being of our children and future generations to come. We all are blessed with talents, aspirations, gifts, and dreams. Knowing these weaknesses and working hard to overcome them brings the promise of success and fulfillment compared with a life of struggles and frustration. I do not see children who are damaged or defective – I see the promise of a future that is clearer due to the insight a diagnosis gives them.

This silent epidemic will require a revolutionary paradigm shift in our social conscience to reduce its skyrocketing trajectory. Such a paradigm shift is necessary to break the bonds of social enmeshment with alcohol – a drug with limited benefit to society outside of its properties as an antiseptic chemical and alternative to fossil fuels. It is now time for society to face the reality that its social drug of choice is aborting human potential silently within the womb. The paradigm must shift toward upstream primary preventive measures which create healthier reproductive health attitudes and lifestyle behaviors.

It is my hope to fuel this new era in women's reproductive health led by the Centers for Disease Control and Prevention, including preconceptional planning and contraception for alcohol consumers. However, in rural and socially disenfranchised communities, it will take more than government ads to turn the tide on this tsunami of lost potential. It is in the spirit of healthier futures that I write – with hope for alcohol-free pregnancies from the preconceptional period through the end of lactation (breast feeding). Informing the public of the fastidiousness of our early developmental stages will be a giant step toward ensuring that future generations have the capacity to fully appreciate the universe in which we live—its forests and flowers, textbooks and tests, microbes and mountains.

Until preconceptional measures are embraced, the rarity of human genius—from Leonardo da Vinci, Anton van Leeuwenhoek, Ludwig van Beethoven, Charles Darwin, and Albert Einstein—may become forever extinct.

Part III

Professional and Parent Guide to ND-PAE

Appendix A

Through the Lens of a Child Psychiatrist – Clinical Insights into Diagnosis and Treatment

> *Individuals with ND-PAE are at higher risk than unexposed groups for anxiety and depressive disorders. Research shows that prenatal alcohol-induced reprogramming of hypothalamic-pituitary-adrenal (HPA) and hypothalamic-pituitary-gonadal (HPG) systems may predispose individuals to depression and anxiety throughout the life course.*
> *— SDR, 2015.*

Nearly every week, I encounter either a new patient or the case of an individual with ND-PAE, whose parents have been told, *"There is no reason to tell the child/adolescent they have the disorder"* or *"The diagnosis does not matter – you don't want your child stigmatized by such a label anyway."* Even a fairly high ranking physician with a government agency that funds research into prenatal alcohol exposure was ambivalent at best about letting her adoptive daughter know that she has the condition. Recently, I was speaking with a public defender working with a 12-year-old girl who was adopted during early childhood and residing in Baltimore, MD at a state residential treatment center. The treatment team's position during an educational management team meeting was that the diagnosis has little bearing on her treatment or services. In most forensic cases I have evaluated, the individual has never been diagnosed with ND-PAE and psychiatrists in the system know very little about working with the children. This leads to a complex array of unhelpful diagnoses that respond poorly to medication and/or other treatments and children who are medicated to the point of being vegetative.

Children who are diagnosed early with ND-PAE have the benefit of their parents learning about their neurodevelopmental challenges, understanding their limitations, and helping them cultivate their strengths and interests. Early diagnosis also helps provide early intervention services, therapy, and supports for the children and their parents. Therapists teach ways to help them understand and deal with frustration, blows to their egos, disappointments, and conflict. Other allied health professionals (speech/language, occupational, and physical therapy) can help with other areas of functioning, such as language, communication, daily living skills, and motor skills. Without early diagnosis and intervention, children with ND-PAE are left to their own primitive coping strategies – and sometimes with parents who are themselves poorly emotionally regulated, doing more harm than good with discipline approaches.

Prior to the psychiatric evaluation, I have the parents complete a childhood screening questionnaire in order to assess the risk of prenatal alcohol and other substance exposure as well as to evaluate early childhood indicators of neurodevelopmental issues. Sample questions are depicted in Appendix B-1. Although these questions are helpful for all children during a psychiatric assessment, regardless of prenatal exposure histories, they can be particularly helpful in understanding the early developmental pathways of kids with ND-PAE. Through this process I evaluate a child's strengths and challenges, opportunities and threats, environmental supports and stressors – in short, creating a strategic plan. Early, accurate diagnosis and providing the proper environment for the child to flourish are equally important to the right medication and treatment plan.

While my approach is not the only way to treat ND-PAE, I believe a paradigm shift toward recognition and treatment of this condition will help individuals live happier, more productive and meaningful lives. First, let's begin with the diagnosis ...

Figure A-1: Sample Developmental History

Pregnancy History:

Is the patient your biological or adopted child? _____

How old was the child's biological mother at the time of birth? _____

Was this an unplanned pregnancy (mistimed or unexpected)? Yes No
  If yes, at what point was the pregnancy diagnosed (i.e., how far along were you when you found out you were pregnant)? _____

During pregnancy with this child did the mother experience any medical/health problems? Yes No If yes, please describe. _____

Did the mother use any prescription/nonprescription medicines or other drugs during the pregnancy? Yes No If yes, please describe. _____

Please list the amount of alcohol (if known) consumed during pregnancy (including beer, wine, liquor; e.g., a glass or two of wine with dinner, a few beers on the weekend, etc.). _____

Might there have been a heavy or binge alcohol exposure prior to pregnancy recognition? Yes No If yes, please describe. _____

Did mother smoke cigarettes during pregnancy? Yes No

If yes, number of packs per day. _____

What other substances were used during the pregnancy (e.g., cocaine, heroin, amphetamines, marijuana, etc.)? _____ How much, how often, and when? _____

Birth History:

  a. How many weeks did pregnancy last? (Check one)
     ☐ Fullterm (38+)   ☐ 34-38   ☐ 32-34   ☐ ≤ 32 ☐ ≥ 42

  b. Child's weight at birth: _____. Birth length: _____.

  c. Were there any complications in the labor or delivery? Yes No
     If yes, please describe. _____

d. Number of days/weeks the child spent in the hospital after birth: _____

e. Were there any medical problems identified at the time of birth or during his/her first year of life? Yes No If yes, please describe. _____

f. Was the child hospitalized for any reason during the first year of life? Yes No If yes, please describe. _____

Caregivers:

Briefly describe childcare arrangements from birth to present:

_____
_____
_____
_____

Developmental Issues:

a. During your child's first six months, was he/she (check all that apply):

☐ An easy baby?	☐ Feeding difficulties?
☐ Enjoyed people?	☐ Unusually sick?
☐ Cuddly?	☐ Strong reaction to light?
☐ Sleep/wake cycle regulated?	☐ Strong reaction to sound?
☐ Difficult to soothe?	☐ Strong reaction to smell?
☐ Irritable?	☐ Strong reaction to textures?
☐ Unusually sleepy?	☐ Strong reaction to touch?
☐ Unusually quiet?	☐ Colicky (reflux or g.i. issues)

b. How old was your child when s/he met the following developmental milestones?

Developmental Milestones	Age (mos/yrs)	Developmental Milestones	Age (mos/yrs)
Smiled at caregiver		Babbled	
Waved "bye-bye"		Understood single words	
Sat by self		Said first word other than "mama" or "dada"	
Crawled		Put two words together	
Walked by self		Spoke in sentences	
Rode a bike		Spoke intelligibly (so others could understand)	
Fully Toilet trained		Dressed self independently	

c. How old was your child when you first became concerned about his/her development (if applicable)? _____

d. What were your initial concerns and who raised these concerns?
_____
_____
_____

e. Has your child received previous neuropsychological, psychological, educational, speech/language, neurological or other developmental testing? Yes No If yes, please list these and your child's age at the time: _____

f. What was the resulting advice/diagnosis given by this individual? _____

Figure A-2. Screening Checklist for Neurodevelopmental Disorder Associated with Prenatal Alcohol Exposure[154]

Please indicate which criteria your patient meets: If so, check & answer here:

A. ____ More than minimal exposure to alcohol at any time during pregnancy (2 drinks or more; including prior to pregnancy recognition) Was this a planned or unplanned pregnancy? ____	☐ maternal self-report ☐ collateral reports ☐ medical or other records: ____
B. Neurocognitive impairment (at least one of the following) ☐ 1. global intellectual impairment ____ ☐ 2. impairment in executive functioning ____ ☐ 3. impairment in learning ____ ☐ 4. impairment in memory ____ ☐ 5. impairment in visual spatial reasoning ____ ☐ 6. attention deficit ____ ☐ 7. impairment in impulse control &/or hyperactivity ____	Information Source: ____ Date of testing: ____ **Or** Documentation of an Individualized Educational Plan for specific learning disabilities ____
C. Impairment in emotional regulation: ☐ 1. impairment in mood regulation (e.g., mood outbursts, anger, aggression) ☐ 2. autonomic arousal (i.e., anxiety, heightened stress response, hypervigilance, sleep problems)	Please describe: ____
D. Deficits in adaptive functioning as manifested in two (or more) of the following, including at least one of (1) or (2): ☐ 1. conceptual skills (i.e., applying academic knowledge to daily life) ☐ 2. social impairment (e.g., speech/language, friendships, social cues, pragmatics, facial expressions) ☐ 3. impairment in practical/daily living skills (self-care, hygiene, household chores, meal preparation)	Information Source: ____ Please describe: ____
E. Sensory and/or motor impairment (i.e., hyper/hyposensitivities to light/sound/touch/taste/temperature, gross and/or fine motor issues, coordination problems).	Information Source: ____ Please describe: ____
F. The onset of the disturbance (symptoms in Criteria B, C, and D) is before 18 years of age.	**Age of onset:** ____
G. The disturbance causes clinically significant distress or impairment in social, occupational, or other important areas of functioning.	Please describe: ____
H. The disturbance is not better explained by the direct physiological effects of another condition.	Please describe: ____

---

[154] This screening tool is based on a slightly modified grouping of criteria compared with the criteria listed in DSM-5. The adaptive functioning domains align with the criteria for intellectual disability.

Figure A-3. Explanation Checklist - Neurodevelopmental Disorder Associated with Prenatal Alcohol Exposure

Please indicate which criteria your patient meets:

A. Is there documentation of the following?
- ☐ <u>more than minimal exposure to alcohol at any time during gestation</u>, including prior to pregnancy recognition. Confirmation of gestational exposure to alcohol obtained from any of the following sources: (circle) 1. maternal self-report of alcohol use in pregnancy; 2. collateral reports, or 3. medical or other records.

B. <u>Neurocognitive impairment, as evidenced by one (or more) of the following</u>: Source/date of neurocognitive testing: _____
- ☐ global intellectual impairment (i.e., IQ of 70 or below, or a standard score of 70 or below on a comprehensive developmental assessment).
- ☐ impairment in executive functioning (e.g., poor planning and organization; difficulty changing strategies or inflexibility;).
- ☐ impairment in learning (e.g., lower academic achievement than expected for intellectual level; requires special education services; specific learning disability)
- ☐ impairment in memory (e.g., problems remembering information learned recently; repeatedly making the same mistakes; difficulty remembering lengthy verbal instructions)
- ☐ impairment in visual spatial reasoning (e.g., disorganized or poorly planned drawings or constructions; problems differentiating left from right; problems aligning numbers in columns)
- ☐ attention deficit (e.g., difficulty encoding new information; difficulty shifting attention; difficulty sustaining mental effort)
- ☐ impairment in impulse control (i.e., difficulty with behavioral inhibition &/or hyperactivity, difficulty waiting turn; difficulty complying with rules; confabulating; taking possessions of others)

C. <u>Impairment in emotional regulation in one (or more) of the following</u>:
- ☐ 1. impairment in mood regulation (e.g., mood lability; negative affect or irritability; frequent behavioral outbursts).
- ☐ 2. autonomic arousal (i.e., anxiety, heightened stress response, hypervigilance, sleep problems)

A. Deficits in social/communication or other areas of <u>adaptive functioning</u> as manifested in two (or more) of the following:
- ☐ 1. conceptual skills (i.e., applying academic knowledge to daily life)
- ☐ 2. communication deficit (e.g., delayed acquisition of language; difficulty understanding spoken language; difficulty using language to express self so that the listener understands). Please describe and list the source of the information: _____
- ☐ social impairment (e.g., delays in speech/language, immature or inappropriate friendships, doesn't get social cues, pragmatics, facial expressions overly friendly with strangers; difficulty reading social cues; difficulty understanding social consequences; acting "too young").
- ☐ impairment in daily living (e.g., delayed toileting, feeding, or bathing; problems following rules of personal safety, self-care, hygiene, household chores, meal preparation; difficulty managing daily schedule).

B. Sensory and/or motor impairment (i.e., hyper/hyposensitivities to light/sound/touch/taste/temperature, poor fine motor development; delayed attainment of gross motor milestones or ongoing deficits in gross motor function; problems in coordination and balance).

E. The onset of the disturbance (symptoms in Criteria B, C, and D) is <u>before 18 years of age</u>. **Yes No Age of onset:** _____

F. The disturbance causes <u>clinically significant distress or impairment in social, occupational, or other important areas of functioning</u>. Yes No
   If yes, please describe: _____

G. <u>The disturbance is not better explained by the direct physiological effects [of another condition]</u> associated with postnatal use of a substance (e.g., medication, alcohol or other drugs), another medical condition (e.g., traumatic brain injury, delirium, dementia), other known teratogens (e.g., Fetal Hydantoin syndrome), a genetic condition (e.g., Williams syndrome, Down syndrome, Cornelia de Lange syndrome), or environmental neglect. Yes No Please explain: _____
_____
_____

Appendix B

Therapeutic & Learning Centers – A Neurodevelopmental Treatment Model

Because effects of prenatal alcohol exposure can mimic a variety of psychiatric disorders, it is important to accurately diagnose their condition in order to develop the most appropriate treatment plan. Often, I speak with psychiatrist colleagues who say they will not change their diagnostic frame from ADHD or autistic disorders to ND-PAE, even in children who clearly meet the Diagnostic and Statistical Manual, Edition 5 (DSM-5) criteria for the disorder. Today, I spoke with a child psychiatrist who said, "There are many environmental causes of psychiatric conditions. The diagnosis is what matters, not the etiology." This was the same position initially taken by the American Psychiatric Association leadership in 1998 when the National Organization on Fetal Alcohol Syndrome and other concerned agencies convened a conference in Washington, DC to discuss the possibility of including the conditions associated with prenatal alcohol exposure in DSM. Why does etiology matter? Won't the treatment be the same?

The perspective of those of us who have thought about, treated, researched and evaluated children with effects of prenatal alcohol exposure is that etiology does matter. Moderate to heavy prenatal alcohol exposure can cause a wide range of deficits in children that are somewhat resistant to standard techniques of treatment. The medications we use and the therapies can be counterproductive, even harmful, if etiology is not taken into consideration. For example, since the children may have underlying cardiac defects, conduction anomalies, or arrhythmias associated with prenatal alcohol exposure, it is especially important to rule out underlying heart conditions for them prior to beginning treatment with stimulants. Additionally, leptomeningeal heterotopias and other brain anomalies caused by prenatal alcohol exposure can be linked with seizure disorders. Medications that lower the seizure threshold can sometimes unmask such conditions (i.e., buproprion). An understanding of etiology also enables astute clinicians to rule out cytochrome P-450 enzyme genetic differences as well as serotonin transporter and receptor genotype – any of which would affect an individual's response to certain medications.

In my clinical practice, I see children and adolescents with multiple, complex overlapping areas of mood instability and anxiety, neurocognitive deficits, social communication problems, and motor/coordination/sensory issues. Often, these young people come with a host of diagnoses (I call it, "Alphabet Soup") and numerous medications, some of them started while they were hospitalized in a state of crisis. One of my patients took herself off the cocktail of pharmaceuticals during a wilderness program after three failed hospitalizations and a series of residential programs over a 6-month period of time. Her father took the photograph of the collection of pills she was taking when she was discharged from the last program for throwing rocks at staff and refusing to comply with their safety protocols.

Additionally, traditional psychotherapy, cognitive and behavioral approaches, and other forms of non-pharmacological treatment can likewise harm rather than help the patient with ND-PAE. Many of the children may have subtle receptive and/or expressive language deficits, nonverbal learning disorders, social pragmatic challenges, or other communication issues impacting their ability to understand traditional psychological approaches to management and treatment. Other children may have auditory processing issues, cognitive or executive functioning problems, or other difficulties interfering with their ability to benefit from traditional behavioral management. Their reactive nervous systems may go into a fight or flight mode, leading them to over-react to parental redirection, constructive criticism, or behavioral consequences. Less threatening approaches to parenting are often more helpful, such as a reward based system, gentle tone and praise, and structured daily scheduling can often minimize triggers and prevent escalation of their behaviors.

In my clinical practice, I have developed a four domain model of ND-PAE depicted in Figure 1 providing a structure to understand the complex array of neurodevelopmental issues more or less affected with prenatal alcohol exposure: emotional regulation (i.e., heightened stress response), social communication, neurocognitive functioning, and motor/coordination/sensory ability. This system is not meant to be a diagnostic tool as it is not modelled directly from the DSM-5 criteria. Instead, it a teaching aide for parents and caregivers and a guide for psychiatrists in neurodevelopmental treatment planning to ensure efficiency and thoroughness in clinical practice. An astute fellow in child psychiatry at the State University of New York - Upstate asked during a Grand Rounds a couple of years ago whether the same Venn diagram could be applied to any neurodevelopmental condition. In my experience, non-exposed children may have one developmental deficit (i.e., sensory integration or fine motor issues) but not several areas of deficiency. The overlapping areas indicate that individuals with prenatal alcohol exposure can have one or more domains of impairment.

The degree of effects depends upon a mother's nutrition status, genetic predisposition, other lifestyle behaviors (e.g., smoking cigarettes, recreational/illicit substance use), stress level, medical issues (i.e., diabetes, thyroid issues), as well as genetic predisposition in the fetus. The timing, duration, frequency, and maternal alcohol concentration also contribute to the range and degree of deficits, with earlier and binge exposures frequently associated with worse outcomes.

Using this approach, individuals suspected of ND-PAE would be referred for the following assessments: neuropsychological testing to rule out neurocognitive issues, speech/language screening to understand their social communication problems, and occupational/physical therapy to identify and treat underlying fine/gross motor deficits and/or other functional issues. Mood regulation and autonomic arousal should ideally be assessed by a board certified child/adolescent psychiatrist and/or a psychologist. This team of professionals works with the individual and her family to develop a comprehensive treatment plan to enhance strengths and support challenges. Assessments should include their interests, abilities, and other areas to build vocational skills and self-esteem through mentoring, apprenticeships, and experiential learning as they mature into adolescence. Using a strengths-based approach, the team also helps the parent and child develop extracurricular and academic activities to offset the rigors of academics.

Why is it that children's brains can be affected in such unique and different ways? A basic description of the nervous system helps guide understanding of how prenatal alcohol's effects lead to such a wide range of outcomes. The nervous system is divided into two parts: central and peripheral. The central nervous system is considered the brain and spinal cord, regulating a variety of bodily functions as well as thoughts, feelings, and behavior. The peripheral system is divided into two parts: voluntary and involuntary. Whereas the voluntary nervous system is primarily under the individual's control (i.e., intentional motor movements and volitional actions), the involuntary system is largely not under the individual's conscious control.

Appendix Sections B-1 to B-4 highlight the neurodevelopmental issues to consider during the evaluation and management process. The chapter, "Nature versus Nurture – Or Both!" discusses how the environment can either improve or worsen the outcome in the child.

Figure B-1: Domains of ND-PAE[155]

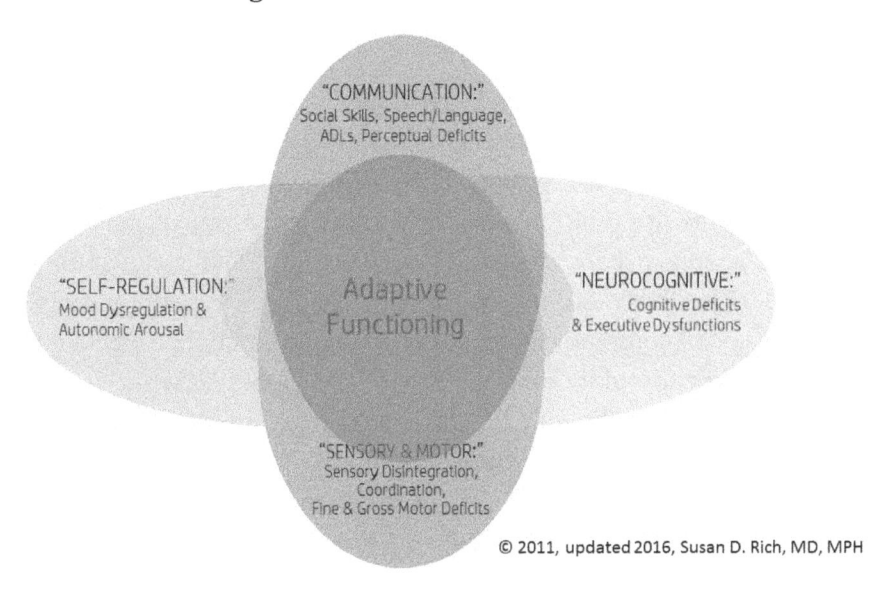

"COMMUNICATION:"
Social Skills, Speech/Language, ADLs, Perceptual Deficits

"SELF-REGULATION:"
Mood Dysregulation & Autonomic Arousal

Adaptive Functioning

"NEUROCOGNITIVE:"
Cognitive Deficits & Executive Dysfunctions

"SENSORY & MOTOR:"
Sensory Disintegration, Coordination, Fine & Gross Motor Deficits

© 2011, updated 2016, Susan D. Rich, MD, MPH

Adaptive Functioning

Adaptive behavior is defined by the American Association on Intellectual and Developmental Disabilities as *"the collection of conceptional, social, and practical skills that are learned and performed by people in their everyday lives.*

- *Conceptual skills—language and literacy; money, time, and number concepts; and self-direction.*
- *Social skills—interpersonal skills, social responsibility, self-esteem, gullibility, naïveté (i.e., wariness), social problem solving, and the ability to follow rules/obey laws and to avoid being victimized.*
- *Practical skills—activities of daily living (personal care), occupational skills, healthcare, travel/transportation, schedules/routines, safety, use of money, use of the telephone.*

---

[155] Kieran D. O'Malley and Susan D. Rich. Clinical Implications of a Link between Fetal Alcohol Spectrum Disorders (FASD) and Autism or Asperger's Disorder – A Neurodevelopmental Frame for Helping Understanding and Management. Chapter 20 in *Recent Advances in Autism Spectrum Disorders - Volume I*, book edited by Michael Fitzgerald, March 6, 2013.

Standardized tests can also determine limitations in adaptive behavior."[156]

Adaptive functioning relates to one's ability to handle common life demands and level of independence compared to others of a similar age and background. In this way, one's adaptive functioning level is affected by all domains listed in Figure 1. The ability to adapt to a number of different people, events, situations, environments, and social settings depends collectively on neurocognition, executive functioning, and intellect; social communication and relatedness; emotional regulation; and motor/sensory abilities. Throughout childhood, adolescence and adulthood, individuals with ND-PAE tend to be immature, gullible, easily influenced by peers, impulsive, and emotionally labile. They often have difficulty managing money, maintaining an apartment, paying bills, making and keeping appointments, and taking care of their own children. Socially, their friendships with typically developing peers are limited, leading them to develop relationships with individuals who are functioning at similarly low levels. Due to expressive and receptive language issues, misperception of social cues, pragmatic problems, and emotional dysregulation, they can appear paranoid, suspicious, or easily triggered into altercations with peers.

The combination of these neurodevelopmental and psychiatric sequelae frequently persist through the life course. Without proper treatment, vocational supports, supportive housing, and mentoring, progressive worsening of these conditions can lead to devastating outcomes and poor prognosis.[157] For these reasons, many are at risk for delinquency, teen pregnancy, alcoholism, substance use disorders, and a variety of other unfortunate outcomes. In this sense, our misguided and ill-informed drinking behaviors have contributed to the social disparities and mental divide that make it so difficult for some people to succeed beyond the underclass of society.

---

[156] http://aaidd.org/intellectual-disability/definition#.VZgvV_1VhBc
[157] Streissguth AP1, O'Malley K. Neuropsychiatric implications and long-term consequences of fetal alcohol spectrum disorders. *Semin Clin Neuropsychiatry.* 2000 Jul;5(3):177-90.

Appendix B-1

EMOTIONAL REGULATION:
MOOD DYSREGULATION & AUTONOMIC AROUSAL

Emotional regulation is controlled by the central nervous system, but can be influenced by the peripheral involuntary nervous system, which is divided into two parts. The sympathetic system exists to help the animal or person conserve resources in order to "fight" or "flee" during episodes of danger. The parasympathetic or "calm" system is active only when the sympathetic system is turned off – lowering the heart rate, blood pressure, body temperature, as well as controlling metabolism, digestion, immunity, and circadian (sleep-wake) cycles.

Magnetic resonance imaging of the brain of adolescents with FAS shows a thin or non-existent corpus collosum—the crescent moon-shaped midline structure connecting the two hemispheres of the brain. Sometimes thinning of the cortical white matter is seen as well as volume reductions in other areas of the brain. Selective reductions in the corpus collosum are similar to those reported in children considered to have attention deficit and hyperactivity disorder (ADHD) and information processing problems. Attention deficits—the inability to stay focused on a task, to follow rules, finish chores, school assignments, or keep commitments—have long been recognized as characteristic of children with FAS. The extent to which recent increases in the incidence of children with ADHD may be attributed to ND-PAE is unclear; however, in my own practice, prenatal alcohol exposure often causes ADHD symptoms to be more difficult to treat (refractory), the child to have paradoxical reactions to medications, and the side effects to be more severe.

Prenatal alcohol exposure is the leading known cause of intellectual disability in the Western World and the leading preventable cause of birth defects. The average child with alcohol-related birth defects has an IQ of 70—but can be classified as intellectually disabled based on adaptive functioning scores—with a wide range of scores from about 50 to 100. In addition to lowered IQ, the spectrum of brain abnormalities caused by alcohol translate into neurodevelopmental problems. Our human cognition—not to be taken lightly—allows us to discover new scientific theories, create symphonies, paint elaborate works of art, build skyscrapers, solve complex math equations, design computers, and understand spiritual truths. Alcohol attenuates these innate, God-given functions.

Children with ND-PAE may be unable to adapt to their surroundings, to accept responsibility, to identify social cues, to demonstrate appropriate behavior, and to bond with peers. Given the nature of their impairments, they lack self-discipline, are short-sighted and impulsive, easily swayed by peers, and overreact to stressors. Described by their parents and caretakers as "Moral Chameleons," it is no wonder that 60% of children with FAS have been incarcerated before the age of 18.

In short, the brain circuits of an infant exposed prenatally to alcohol is something like the wiring of an 1812 farm house jerry rigged for electricity in the 1930's. Many of the circuits will get easily overloaded, tripping the breaker and knocking the power off momentarily or overheating and causing an electrical fire. Alcohol weakens the integrity of neurons – stripping them of their protective insulating cover called "myelin," just as bleach soaks into and weakens the fibers of clothing. This causes the cells to transmit their messages in a fragmented or haphazard way – often ending up sending signals to a different part of the brain than the intended "trigger." The long term home owner of such a temperamental house knows all too well that certain switches or outlets cannot all be used at the same time without risk of a momentary power failure. In such an old house, there will be the occasional mouse that chews through the protective sheath covering wires in the walls. When a switch is later turned on, those frayed wires may send sparks flying, leading to smoldering and sometimes a house fire. Individuals with these faulty circuits (and their caregivers or family members) are often caught off guard as a seemingly benign trigger flips their demeanor into a completely enraged state. In other cases, it is like a burned out light bulb with intermittent power going to it. For others, it is like having the light bulb in the wrong place (in a dresser drawer instead of in the light fixture) and power re-routed to the furnace. The brain's amygdala is like the emotional rheostat – the "red mark" in the thermostat that tells the furnace to turn off when the room has reached the desired temperature (or to turn on when it has gotten too cool). Because this midline structure can be damaged in individuals with ND-PAE, they are more easily triggered into a "fight or flight" mode. Early life trauma (witnessing or experiencing abuse) further compromises this primitive structure of the brainstem, increasing its size from that of a pea to an almond. These individuals find themselves in a perpetual cycle of heightened stress response with few internal resources to self-regulate or adapt to environmental triggers.

Other areas of the brain, including the prefrontal cortex and the sulci and gyri of the neocortex are affected in such a way to cause mild intellectual impairment, issues distinguishing consequences and understanding right from wrong. Whereas genetic disorders such as Down's syndrome tend to cause global intellectual impairment in the mild to moderate range, prenatal alcohol exposure tends to be associated with less marked impairment in full scale IQ and more significant damage in discrete areas. The individual tends to have relative strengths and a number of challenges – such as in social reciprocity, pragmatics, receptive language, processing speed, working memory, and other measures of "executive functioning." Like latent syphyllis that spreads gradually through the neurons, eventually causing mental decline and disabling neuronal degeneration, the evidence on human circuits is not easily recognizable at birth and therefore more difficult to track compared with physical birth defects caused by other chemicals. Such deficits can be profound – leaving individuals functioning at the level of 1% of the population on some measures.

The evolutionarily protective sympathetic system is triggered during times of stress to warn the person (or animal) about life-threatening situations, such as predators, fire, and environmental catastrophe. The aggressive sympathetic system over-rides the relatively more passive parasympathetic (tend and mend), shutting down the nonessential organ systems (stomach, gut, immune glands, pancreas) when danger is near. The eyes, ears, nose, or other senses send a signal to the brain, pushing the "panic button" (amygdala). This turns on the fear/stress response (the "sympathetic" reaction) – signaling release of a cascade of hormones and neurotransmitters through a feedback loop in the brain known as the hypothalamic-pituitary-adrenal (HPA) axis. This "fight or flight" mechanism increases the heart rate, breathing rate, and blood pressure in order to help the animal or person escape the dangerous situation.

Why children with ND-PAE experience a physiological heightened stress response relates to direct damage to the brain "hard wiring" (neurons), as well as hormonal disruptions in the HPA axis. Alcohol consumed prior to pregnancy recognition (e.g., during the third to fourth week after conception) affects the way that the baby's brain cells migrate from their origin to their destination. Changes in neuronal wiring, brain architecture, and molecular mechanisms (receptors for and production of serotonin and other neurotransmitter systems) thus exacerbate anxiety and mood disorders. Alcohol's silent effects on the amygdala and HPA axis – a primitive but powerfully protective alarm system, contribute to an over-reaction to environmental stressors. This key system involved in the stress response is sensitive to "prenatal programming" by alcohol. Alcohol consumed during pregnancy crosses the placenta and affects endocrine organ development, altering the interaction between the mother's and fetus's hormone systems. These changes affect prenatal metabolism, physiology, and hormone functions.[158] Prenatal alcohol exposure thus upsets the balance between "parasympathetic" and

---

[158] Weinberg J, Sliwowska JH, Lan N, Hellemans KG. Prenatal alcohol exposure: foetal programming, the hypothalamic-pituitary-adrenal axis and sex differences in outcome. *Journal of Neuroendocrinology*. 2008 Apr;20(4):470-88.

"sympathetic" control.[159] The affected individual goes into sympathetic overdrive (the "fight or flight" reaction) more easily.

My patients have taught me that these challenges with the neuronal circuitry in ND-PAE make it difficult to "dial down" responses to stressors, fear, excitement, and anger. For this reason, individuals with ND-PAE are more at risk for stress-induced illnesses.[160] For example, if they also live in stressful environments (i.e., ghettos, barrios, abusive homes), they will be even more susceptible to life stressors than typically developing (non-alcohol exposed) children (see also chapter on "Nature versus Nurture – or Both!").

An important molecule in the HPA Axis is Proopiomelanocortin (POMC) – a large "precursor" molecule produced in the pituitary, hypothalamus, brainstem and skin cells. POMC in the brain is broken down (cleaved) into a variety of chemicals, ranging from opioids to melanin and corticosteroids. Functionally, these diverse chemicals have roles in regulating appetite, sexual behavior, immune function, and secretion of melanin. Alterations in the gene due to mutations have been linked to obesity and adrenal insufficiency. In the skin cells (epithelium), placenta, and hypothalamus, the POMC peptide is cut (cleaved) into many "peptides" (small molecules) with roles in pain, immune function, energy regulation, and stimulation of skin cells to produce the pigment, melanin. POMC stimulates ACTH, which promotes glucocorticosteroid (GC) release.

> *GCs reduce inflammatory responses through an immunosuppressive action, and stimulate the sympathetic "fight or flight" response giving a rapid, temporary boost to an organism responding to an environmental threat. The autonomic nervous system complements the stress response, either through "fight or flight" (mentioned above), or by "tend and mend," an opposing process through the*

---

[159] Epigenetic mechanisms (i.e., methylation of DNA in "germ cells" – sperm or eggs) also influence the way that genes are turned on or off ("expressed"). These effects account for some of the predisposition that people have to alcoholism and substance use disorders as well as mood and anxiety disorders.

[160] Weinberg J, Sliwowska JH, Lan N, Hellemans KG. Prenatal alcohol exposure: foetal programming, the hypothalamic-pituitary-adrenal axis and sex differences in outcome. *Journal of Neuroendocrinology.* 2008 Apr;20(4):470-88.

parasympathetic nervous system. ACTH also stimulates the production of catecholamines (CATs) from the adrenal glands. Epinepherine, also known as adrenaline, and its counterpart, norepinephrine (noradrenaline) are CATs released by the adrenal medulla that activate the sympathetic stress response, leading to many of the common physiological symptoms of stress, such as sweating, dry mouth, and a rapid heartbeat.

Though the sympathetic stress response increases survival, it comes at a significant cost in terms of metabolism, the immune system, digestion, and other physiological processes, and cannot be maintained indefinitely. POMC neurons, located in the hypothalamus, are critical for bringing stress homeostasis. The perturbation of the HPA axis is a direct cause of many of the symptoms associated with FASD including deficient stress response, depression, anxiety, and impaired immunity ... At the molecular level, many intertwined causal factors contribute to FASD leading to the varied impacts seen among those who suffer. Among the factors, FASD is intimately tied to hyper-stress response and anxiety disorders that are connected to the dysregulation of hypothalamic-pituitary-adrenal (HPA) axis functions. [161]

Children with heavy prenatal alcohol exposure also may have cellular-level alterations in the amygdala – the brain's "sensitivity valve" to dampen or modulate their reactivity to stressors – that is, a faulty "emotional rheostat" (like the thermostat's temperature regulator). Animal studies have confirmed what is seen in humans – that anxiety-like behavior is associated with structural changes in the amygdala in rats prenatally exposed to even light amounts of alcohol. There is evidence that low dose prenatal alcohol exposure causes changes in the brain's emotional rheostat at the level of the "dendrite" (extensions of the brain cell or neuron, like the branches on a tree).

"This study is the first to link increases in anxiety like behavior to structural changes within the basolateral amygdala in a model of prenatal ethanol exposure. In addition, this study

---

[161] Mead EA and Sarkar DK. Fetal alcohol spectrum disorders and their transmission through genetic and epigenetic mechanisms. *Frontiers of Genetics.* 2014; 5: 154.

> has shown that exposure to even a relatively small amount of alcohol during development leads to long term alterations in anxiety-like behavior." [162]

Evidence from animal studies has been confirmed in human populations – indicating that even low to moderate use of alcohol affects the HPA (stress response) axis.

> "[Prenatal drinking days (PDD)] from conception to pregnancy recognition was related to increases in cortisol reactivity, elevated heart rate, and negative affect in their infants. The effects of PAE on infant responsiveness were significant after controlling for the effects of maternal depression and annual income. In addition, the effects of PAE on cortisol reactivity differed for boys and girls ... Greater PAE was related to greater activation of stress response systems. Our findings suggest that PAE affects the development of infant stress systems and that these effects differ in boys and girls. This work supports the possibility that PAE is related to alterations in infant stress systems, which could underlie problems in cognitive and social-emotional functioning that are common among persons exposed prenatally to alcohol." [163]

Additionally, some studies suggest that the changes are epigenetic in origin and may lead to changes that are inherited by offspring. For example, if a male's sperm are affected by his own heavy alcohol exposure, his offspring are more likely to have mood and anxiety problems. A recent landmark study in this area demonstrates these effects:

> *The HPA axis is a major neuroendocrine system with pivotal physiological functions and mode of regulation. This system has been shown to be perturbed by prenatal alcohol exposure ... is long-lasting and is linked to molecular, neurophysiological, and behavioral changes in exposed*

---

[162] Cullen CL, Burne THJ, Lavidis NA, Moritz KM. Low Dose Prenatal Ethanol Exposure Induces Anxiety-Like Behaviour and Alters Dendritic Morphology in the Basolateral Amygdala of Rat Offspring. *PLoS ONE* 8(1): e54924. doi: 10.1371/journal.pone.0054924

[163] Haley DW, Handmaker NS, Lowe J. Infant stress reactivity and prenatal alcohol exposure. *Alcohol Clin Exp Res.* 2006 Dec;30(12):2055-64.

individuals. ... induced by epigenetic mechanisms ... This developmental programming ... induced a hyper-response to stress in adulthood. These long-lasting epigenetic changes influenced subsequent generations via the male germline ... was reversed in adulthood with the application of the inhibitors of DNA methylation or histone modifications. Thus, prenatal environmental influences, such as alcohol exposure, could epigenetically modulate ... neuronal circuits and function to shape adult behavioral patterns. Identifying specific epigenetic factors in hypothalamic ... neurons that are modulated by fetal alcohol and target [the specific] gene could be potentially useful for the development of new therapeutic approaches to treat stress-related diseases in patients with fetal alcohol spectrum disorders.[164]

Prenatal alcohol exposure (PAE) also decreases the number of serotonin (5-HT) neurons in the brainstem of males and females in animal studies, with ovarian hormones having a protective effect.[165] This research indicates that seratonin dysfunction may account for the higher rates of depression/anxiety disorders in individuals with ND-PAE.

Often, these effects of alcohol on the central nervous system (CNS) produce a highly mood dysregulated child, having random or easily provoked episodes of frustration, irritability, aggression, and anger. Infants and toddlers with ND-PAE can present with Regulatory Disorder Type I, II, or III.[166] This may lead to infants and toddlers seeming to be easily agitated, over-stimulated, and hyper-aroused. Like a broken, miswired or disconnected thermoregulator on a household furnace, their brain lacks the proper "rheostat" to appropriately adjust the maximum emotional output appropriate to a given stimulus. This 'faulty rheostat' in the brain leads to an inability to turn down the dial of their emotional furnace. For some, this may be due to the amygdala being miswired to

---

[164] Bekdash R, Zhang C, Sarkar D. Fetal Alcohol Programming of Hypothalamic Proopiomelanocortin System by Epigenetic Mechanisms and Later Life Vulnerability to Stress. *Alcoholism: Clinical and Experimental Research;* Volume 38, Issue 9, pages 2323–2330, September 2014.

[165] Sliwowska JH1, Song HJ, Bodnar T, Weinberg J. Prenatal alcohol exposure results in long-term serotonin neuron deficits in female rats: modulatory role of ovarian steroids. *Endocrinology.* 2014 Jul;155(7):2578-88.

[166] *Diagnostic Classification of Mental Health and Developmental Disorders of Infancy and Early Childhood,* Revised (DC:0-3R), 2005.

the thalmus and other brain circuits, whereas other individuals may have a faulty amygdala. This leads to impulsive aggression and physically lashing out during stressful or emotionally charged situations.

Like a disconnected or corroded rheostat in a thermometer, the prenatal alcohol exposed brain can be left unable to modulate its "emotional temperature" – leading to drastic and rapid mood fluctuations provoked by minimal stimuli. Many are prone to mood dysregulation that mimics bipolar disorder, intermittent explosive outbursts, and episodes of rage triggered by the slightest insult, sideways glance, or annoyance. As a result, they are more vulnerable to emotional over reaction, poor frustration tolerance, hypersensitivity to criticism, suspicion about both positive and negative stimuli, and being overwhelmed by minimal life stressors. Like the unpredictability of the Earth's weather in the face of Global Warming, these emotional storms may not be as easily predicted as in a normally developing brain. The emotions may shift as intensely as a tsunami caused by turbulence of underwater shifting tectonic plates or a series of powerful tornadoes coming from super cells in the Midwest.

Such altered "arousal patterns" are associated with difficulties settling in the evening to sleep. Altered or disrupted sleep has been described by many parents and caregivers of children with ND-PAE and documented in journal articles. At the same time, few studies have been conducted to tease out the differences between typically developing children's sleep disorders and those of children with ND-PAE.[167] In animal models, prenatal alcohol affects the circadian rhythm in the hypothalamus by changing the metabolism in POMC-producing neurons as well as the expression of "clock regulatory genes."[168]

---

[167] Chen ML, Olson HC, Picciano JF, Starr JR, and Owens J. Sleep Problems in Children with Fetal Alcohol Spectrum Disorders. *J Clin Sleep Med.* 2012 Aug 15; 8(4): 421–429. doi: 10.5664/jcsm.2038

[168] Agapito MA, Zhang C, Murugan S, Sarkar DK. Fetal alcohol exposure disrupts metabolic signaling in hypothalamic proopiomelanocortin neurons via a circadian mechanism in male mice.

Appendix B-2

NEUROCOGNITIVE:
COGNITIVE AND EXECUTIVE DYSFUNCTIONS

Dr. Anne Streissguth published a study showing that children born to parents who are social drinkers have an average of 10 points lower intellectual quotient than children of non-drinkers. Infant and child death is at the extreme end of the continuum of prenatal alcohol exposure. In my clinical experience, children with ND-PAE often present with higher degrees of cognitive and executive function deficits than children with garden-variety (i.e., genetically inherited) Attention Deficit Hyperactivity Disorder (ADHD). Case controlled studies have attempted to tease out the differences in these clinically similar groups.

As far back as the 1960s, the term "minimal brain dysfunction" (MBD) was used to describe children with symptoms similar to those who might be diagnosed with attention deficit hyperactivity disorder (ADHD) today. By the mid-1970s, MBD was known to be a neurodevelopmental disorder affecting nearly 20% of school children. Sir Michael Rutter initially described MBD as subclinical damage on a continuum of gross traumatic brain damage leading to psychological sequelae versus a genetically determined hyperkinetic syndrome.[169]

The following symptoms characterized MBD:

---

[169] Rutter M. Syndromes attributed to "minimal brain dysfunction" in childhood. *The American Journal of Psychiatry*, Vol 139(1), Jan 1982, 21-33.

- ✓ immature development of activity-control centers, emotional regulation, and behavioral disinhibition.
- ✓ academic performance: poor concentration and attention, difficulty blending auditory and visual processing for language performance; academic skill deficits in reading, writing, and math; scatter in testing scores with a marked discrepancy between evaluated potential and actual classroom achievement.
- ✓ "soft signs" of neurologic impairment evidenced on physical examination: primitive or asymmetric reflexes, tactile or other sensory deficits, and/or motor abnormalities.

Early on, clinicians understood that early diagnosis and intervention could improve the outcome for children with MBD, as well as children who eventually were diagnosed with ADHD (and today, ND-PAE). The following parent guidance was recommended and is appropriate for children with ND-PAE (as with other forms of "MBD:"

- ➤ decreased academic pressures: this means it is less important for the child to be able to comprehend or complete algebraic equations in elementary school when much of the time even simple math facts are difficult for them to learn. It is more important to focus on daily life skills, manners, having good eye contact, and understanding emotions.
- ➤ limited social pressures: in other words, reducing the stress produced by crowded environments, parents' expectations of or insistence on managing multiple friendships, and pushing them to "grow up too fast." Children with ND-PAE (as with MBD) are immature socially and will need more time to develop the areas of their brain to interact at the age of their typically developing peers.
- ➤ appropriate (not harsh or unfair) consequences: parents should use consequences for behaviors as learning points to help encourage positive behaviors and minimize negative or unwanted behavior and minimize ridicule (demoralization/demeaning comments) from parents, coaches, and teachers.

- the educational environment: understanding and capitalizing on the child's strengths while supporting areas of relative weakness will ensure that the child is able to develop a healthy self-esteem, build a fund of knowledge in a particular area of interest, and begin to grow into a confident, happy, and hopeful adolescent.
- medications: strategic use of medications to improve attention and "wake up" the sleepy secretary (hypoactive prefrontal cortex) can be helpful but must be considered in a risk:benefit ratio (weighing the side effects such as appetite suppression and tics with the relative improvement in symptoms).
- Therapy: to prevent severe personality disorders (antisocial, borderline, etc.) and delinquent behavior, as well as psychiatric illnesses, therapy should begin at an early age and include parent guidance. It is possible that a large percentage of the children once diagnosed with MBD would fit the diagnostic criteria for ND-PAE since they have overlapping diagnostic criteria.

Due to a hypoaroused, misconnected, or disconnected prefrontal cortex, individuals with ND-PAE have a range of deficits in cognitive areas, including but not limited to: difficulty with executive functioning (organization, concentration, auditory processing, processing speed, visual integration, working memory, problem solving, attention, impulse control, etc.); deficits in IQ often compared with their biological parents; mathematics disorders, reading disorders (e.g., dyslexia), spelling issues, and other learning disabilities. These disruptions in cognitive functioning often lead to a failure to understand consequences, poor judgment, and limited insight into the origin or impact of one's behaviors. This may lead to significant and debilitating deficits in functional abilities. Rather than thinking through actions, the person may act impulsively in a primitive manner (as though driven by primal instinct rather than intellect).

In an article appearing in the November 2012 edition of *Attention*, the magazine of the Children and Adults with Attention Deficit Disorder (CHADD) advocacy and education organization, I discussed my perspective of the difference between intellect and executive functions in lay terms. Below is the article which mostly applies to children with ADHD in general (e.g., genetically-acquired) but can be helpful for understanding executive dysfunctions resulting from prenatal alcohol exposure. Within the passage are descriptions of helpful strategies for compensating for the sleepy secretary.

Executive Functions: the Brain's Secretary[170]

Neuropsychologists describe Executive Functions (EFs) as initiation, attention, working memory, organization, processing speed, filtering information, and a variety of many other highly technical terms. Metaphorically, the Executive Functions (EFs) represent the "secretary" or "executive assistant" to the brain's "Chief Executive Officer" (CEO), or true Intellect. [For younger children, the "school secretary" and "principal" may be described instead.] Breaking it down for patients and their families, the analogy of a secretary versus CEO depicts differences between the EFs and the general intelligence. Vitally important to support the "boss" of the brain - representing abilities in higher level problem solving, critical thinking, abstract reasoning, judging, understanding consequences, and predicting outcomes – the EFs cannot replace the general intellect.

In this scenario, secretarial functions include filing, organizing the office, scheduling appointments, sorting the mail, and other tasks that make the CEO's job a lot easier. Secretaries remember who just called on the phone long enough (i.e., working memory) to direct the call to the correct person in an ancillary department (i.e., processing) and to screen both visitors and callers (i.e., extraneous noises, voices, information, thoughts) from reaching the CEO when s/he is busy. A good secretary is an efficient note-taker (i.e., uses short-hand in order to capture important information) and an acceptable editor (i.e., corrects typo's and punctuation errors, dots the i's and crosses the t's). Although s/he may not grasp the ramifications of a multi-billion-dollar project, s/he is adept at looking at the details of a contract without being overwhelmed by the minutia.

---

[170] Rich SD. Excerpted and adapted from Executive Dysfunctions & the Sleepy Secretary of ADHD, *Attention Magazine*. August 2012.

A wonderfully gifted secretary is productive even when multi-tasking – while covering the phone line and front desk, s/he types a dictation or creates a memo. S/he is able to prioritize the work on her desk in order to know what must get done ASAP and what can wait until after lunch or tomorrow. Secretaries make "to do" lists and follow them until completion, not too proud to make coffee if it puts the boss in a better mood in the morning, and shows up on time, willing to work, and motivated to get any job done. S/he finds value in the work for the sake of having a job, is not too egotistical to make her boss look good, and understands that hard work and perseverance are the keys to success in overcoming challenges. Responsible for "housekeeping" functions of the brain, a fabulously poised secretary often makes the boss look efficient, organized, productive, task-oriented and on time – in turn, improving the boss's performance.

The Sleepy Secretary

When children (or adults) have executive functioning issues, the part of the brain in an area of the prefrontal cortex (just behind the eyebrows and forehead) is hypoactive or "sleepy." Using this analogy, a person with ADHD and/or executive functioning issues has a "sleepy secretary" - one that was out partying most of the night and came into work with little sleep and no coffee. Kids with "high engines" or hyperactivity are innately wired to rouse their sleepy secretary by movement (e.g., foot tapping, fidgeting, doodling, wiggling, standing up, or otherwise being in perpetual motion). This allows them to stay more alert and attentive despite the drowsy secretary. Impulsivity is the child's way of being an active participant in a discussion, classroom, family event, social activity, or other occasion while not fully connected to the "rhythm" of the other people they are around. They tend to blurt out answers, butt into conversations, ask inappropriate or off-topic questions, and jump into a situation before thinking about the consequences. Kids with sleepy secretaries are often difficult to waken in the morning, leading to power struggles with their parents due to being late to school and grumpy at home.

The Burnt Out Boss

A complication to the sleepy secretary is the sleepy boss. For kids who aren't getting good sleep, their "CEO" or whole brain also comes to work sleepy - they stayed too late at the same office party their secretary closed down the night before. It turns out that many kids with ADHD don't sleep well because their brains secrete melatonin later in the evening than those without ADHD symptoms. In these cases, adrenaline then kicks in around the time they should be going to bed to hyper-arouse and re-energize the "low engine," over-riding the effects of melatonin even when it is finally released. Not only is the secretary going to be less alert and less productive, but the boss will also be functioning at a minimal level of productivity. S/he will be unable to compensate for the sleepy secretary if they too are less awake and alert. Chronic sleep deprivation in these kids may seem a lot like going to work or school with a hangover or after a "black out" the night before.

The Burnt Out Secretary

EF issues and ADHD commonly lead to children being burnt out, sleep deprived, and depressed from working their CEO into exhaustion. In these cases, the young women are eager to please, academically driven, and motivated to excel. Their anxiety propels them toward perfectionism, and their intellect is able to compensate somewhat for the hypoaroused prefrontal cortex. Since the secretary is less efficient and not doing her job promptly and on time, the CEO then steps in to answer the phone, schedule meetings, type memo's and file the paperwork. These are all tasks the CEO can do but makes his/her job more difficult, slowing down his/her productivity and decreasing optimal functioning.

Ultimately, chronic sleep deprivation results from the CEO forcing them both to "burn the midnight oil" in order to make up for inefficiency and procrastination earlier in the day. On the other hand, even the most talented, self-motivated secretary cannot do the CEO's job - making huge business decisions, consolidating massive amounts of information, recognizing the bigger picture, and creating strategies for improving the company's assets. The other caveat is that, within a single person's brain, there is no way to "fire" one's secretary any more than a CEO would "fire" his mother if she worked for him. So, the highly intelligent person with faulty EFs begins to recognize trouble when the expectations and academic demands of school begin to exceed the CEO's ability to do his/her job plus the secretary's. Very often, this occurs in late middle school to high school or the first few years of college.

The Medicated Secretary

Frequently, problematic behaviors in school related to hyperactivity and inattention leads to children with ND-PAE being identified in preschool when use of medication can affect appetite, growth, and stature. Alternatives to stimulants, such as guanfecine and clonidine decrease the "externalizing behaviors," allowing the child to participate in quiet learning activities such as circle time and learning centers. Academic accommodations, such as small classroom settings, preferential seating close to the teacher, gentle prompting, and extended time for project completion can support the child's attention, eliminating the need for a stimulant.

Stimulant medication serves to artificially arouse the hypoactive prefrontal cortex by acting like a cup of coffee for the sleepy secretary. Many parents, believing their child has garden-variety ADHD are referred to a child psychiatrist by the pediatrician, who may already have tried stimulant medication. Children with ND-PAE often experience side effects from stimulant medications, often paradoxically worsening hyperactivity and impulsivity or leading to anxiety, depression, or symptoms of other co-occurring disorders. Even in typically developing children, stimulant medications can exacerbate underlying anxiety disorders, leading to obsessions, compulsions, crying episodes, poor frustration tolerance, and even motor or vocal tics. Stimulants can also trigger changes in mood, irritability, sleep problems, poor appetite, lethargy, and "low engines." Psychotic episodes are infrequently triggered by rapidly increasing the dose of a stimulant as well as restarting at a previously therapeutic dose after months off a stimulant.

The Meditating Secretary

Meditation and medication have a lot more in common than the first glance. Often I remark that they only have one letter different. Mindlessness as a goal of meditation is the act of letting go of all extraneous thought except what the teacher's voice is saying, what one is reading, or the conversation one is having with a friend. In order to actively connect to the moment, one can focus on his/her breath while letting go of all thoughts. In this analogy, a person is taught to fill his/her head with air like a balloon while breathing in deeply through the nose, collect all thoughts/worries/discomfort within the breath, then breathe out through the mouth – imagining letting all the thoughts/worries/discomforts expel with the breath. The person is taught to actively notice the thoughts/worries/discomfort but to let go of them gently with each breath.

Over time, this method of meditation can improve focus, concentration, attention, efficiency, productivity, and performance. It is the technique used by highly trained athletes to "get in the zone," by actors and politicians to overcome stage fright, by patients with high blood pressure and heart disease to lower their stress levels, and by yogi's and magicians to control their autonomic functions in order to accomplish miraculous, death defying feats. Children with ADHD and EF issues, as well as anxiety and other emotional issues, can train themselves to meditate in order to keep their secretaries focused, alert, actively engaged in learning, and slow down their engine long enough to think twice about what they want to say in class or the answer they want to choose on a test. It is a way in which we as humans can tune into our own unique rhythm that brings us in closer harmony with those around us.

The Accommodated Secretary

Highly intelligent, hard-working kids can adapt and compensate for mild to moderate executive functioning issues by training their secretary to sleep well, show up on time, stay alert, create lists and complete the items, and drill, drill, drill the information into their brains. They learn other important "coping skills" such as organizational strategies (i.e., using color-coded binders and recopied notes), reminders, calendars, phone alarms for appointments, and structure, structure, structure for staying on task. Making one's bed before leaving the house is a way to feel more productive every morning and less distracted when coming home to do school work. Simply put, a bedroom looks much tidier if the bed is made and the clothes are put in a laundry bin. Other ways of organizing include putting shelves in the closet with odds and ends that would otherwise clutter the room – putting papers and mementos in labeled shoe boxes that can be cleaned out and "purged" periodically.

A Child Psychiatrist's Approach

My approach is to first get the child or adolescent sleeping the adequate number of hours required for their age. Most parents are unaware that post pubescent youngsters still need about 9.25 hours of sleep. Children in the "tween" stages (10-13) need around 9.5 – 10 hours. It still baffles me when parents seem confused as to why their elementary to high school aged children have meltdowns, irritability, and two-year-old temper tantrums. They seem shocked to learn that the child getting 8 hours or less of sleep per night is grumpy because they aren't sleeping enough – or are napping, which serves to worsen sleep dysregulation. Normally, I suggest a 30 day trial of my strategy of getting adequate sleep, "unplugging" oneself from electronics for 2 hours consistently after school to focus on homework and studying, and avoiding napping which can further exacerbate the sleep cycle dysregulation. Once the patient is an adequate amount of consolidated sleep for their age, they often require less medication, respond better to accommodations, and are more apt to complete their homework in a timely manner.

There is evidence that children encouraged by statements like *"you are really smart, you should be able to do it"* perform more poorly on standardized testing than their matched counterparts who are praised for their determination and "stick-to-it-iveness." Like my grandmother always told me, *"There is nothing you can't do if you put your mind to it!"* My motto is, *"What separates a smart person from a successful student is hard work and perseverance."* I encourage my patients to minimize distractions of electronics and social media, discipline him/herself to study efficiently (using learned tools and coping strategies), and reframing homework completion as satisfying productivity instead of "busy work." I encourage parents of children and adolescents to avoid micromanaging homework in order to foster independent learning in the child, and to praise the child for diligence, fastidiousness (e.g., making his/her bed in the morning), and the accomplishment of effort rather than grades.

Distractions from Career Goals with ADHD and EF Issues

My patients often hear me say, "In times of catastrophe, I want as many people with ADHD on my side to solve the problem – they act quickly, run on adrenaline, and think outside the box – all great qualities of emergency response teams and entrepreneurs!" Truth be told, many kids with ADHD in my practice are children of entrepreneurs and emergency room physicians. During discussions of family history, the successful parent with ADHD will frequently admit to having taken medication in the past or currently. Genetically speaking, the apple doesn't fall far from the tree! Twenty to thirty years ago, it was much easier for the intelligent student to compensate for a "sleepy secretary." There were certainly far fewer distractions – television, cell phones, video games, social media, and other forms of electronics that feed procrastination, loss of productivity, and inefficiency in even the most intelligent adolescents. Smart, hard-working students could more easily overcome EF issues and ADHD by determination and postponing gratification. The instantaneous gratification of cell phones, video games, instant messaging, texting, video chatting, and other forms of media limits one's drive and motivation for careers requiring years of intensive study. Helping youngsters find their hidden passion, meaning in life, and purpose on the planet is always a goal of my work with patients with EF issues and/or ADHD – to motivate, inspire, encourage, and believe they can reach their realistic goals in life.

In that spirit, the "sleepy secretary" analogy serves to promote the philosophy that a belief in oneself, working hard, and persevering despite academic adversity diminishes moments of frustration and can help improve one's self esteem, motivation and confidence.

Appendix B-3

Social Communication:
Language, Social Skills, & Perceptual Deficits

A variety of speech/language and related socialization disabilities can also be seen in many individuals with ND-PAE. It is important to understand that prenatal alcohol-induced organic brain damage underpins the language deficits. The misuse of language integral to social cognition and communication are quite common problems in adolescents or young adults with ND-PAE. At times, these patients are misdiagnosed with Autistic Spectrum Disorder or Asperger's Disorder.[171] The term "social communication disorder" better fits this population. This does not preclude the fact that medication may engender a positive effect on language functioning, and specifically social communication. Individuals with ND-PAE suffer from indiscriminate or immature behaviors (e.g., telling off-color jokes in the classroom or workplace, telling a person what they think of them even if it is inappropriate and negative). These behavior problems range from silly or irritating socially inappropriate behaviors to overtly aggressive and sometimes risky behaviors. Severe social functioning problems may result in lack of long term friendships, being labeled by peers as "weird" or "odd," and/or appearing withdrawn, socially-isolated, and avoidant.

---

[171] Kieran D. O'Malley and Susan D. Rich. Clinical Implications of a Link Between Fetal Alcohol Spectrum Disorders (FASD) and Autism or Asperger's Disorder – A Neurodevelopmental Frame for Helping Understanding and Management. Chapter 20 in *Recent Advances in Autism Spectrum Disorders - Volume I*, book edited by Michael Fitzgerald, March 6, 2013.

Adolescents or adults who misinterpret social situations may appear to have paranoid behaviors, such as over-reactions to the tone of someone's voice or an otherwise harmless look in their direction. In fact, brain wiring anomalies causing a condition known as "alexithymia" – difficulty experiencing one's own emotions or the emotional world of another person – make them seem callous and unemotional. Because of misunderstanding nonverbal cues, misreading facial expressions, and/or misconstruing what someone says, they may end up in altercations or other compromising social mishaps. At times, ND-PAE may lead to socially indiscriminate behaviors (i.e., individuals engaging in early or promiscuous sexual activity, gang membership, and peer pressure). Because they are easily led and influenced by peers, they are most often the co-defendant "holding the bag" in criminal cases.

Individuals with distortions in social communication and perception due to prenatal alcohol-induced brain wiring differences are prone to diagnoses of personality disorders without appreciation of the etiology of the underlying brain damage. For example, many may appear callous and unemotional due to lack of appreciation of their emotions and that of others. This may lead to a diagnosis of antisocial personality disorder. Others may misperceive social cues, have thought distortions, appear odd and eccentric, and have social awkwardness causing clinicians to think they have schizoid or schizotypal traits. Still others manifest these traits combined with vacillating extreme emotional states (due to mood dysregulation and autonomic arousal) concrete, black and white thinking about social misperceptions then receive the diagnosis of borderline personality disorder.

Appendix B-4

SENSORY AND MOTOR:
SENSORY PROCESSING, COORDINATION & MOTOR ISSUES

Many of my patients with ND-PAE have sensory integration (also known as "sensory processing") issues, including hypo- or hyper-sensitivities to noise, touch, proprioceptive stimuli, smells, tastes, and/or light. As infants, these and other issues (e.g., disruptions in the circadian cycle) lead to sleep/wake problems. They are often difficult to soothe, may not seem to enjoy their caregivers (not cuddly), and can suffer from a range of other regulatory problems. As toddlers and young children, they frequently are sensitive to environmental sounds, lights, fans, and may be easily irritated by loud voices or music. Attending birthday parties, eating in crowded cafeterias, or participating in a noisy classroom can be difficult to manage. Older children, adolescents and adults may cope by avoiding or over-reacting in situations or environments which provoke their sensitivities. In general, transitions are difficult for them, they require intensive one-on-one adult attention, are unable to self-soothe easily, and have difficulty playing independently (but can be taught to do so as a method of self-regulation).

Those with hyposensitivity to light or deep touch may seek out tactile stimulation or may not feel injured when they have a cut requiring stitches or a broken bone. Alternatively, those with hypersensitivity to light touch may be averse to even their parent or caregiver touching them affectionately on the arm or head. I've known patients with ND-PAE who are easily affected by the slightest injury, screaming and writhing in pain when they have the smallest scrape from the playground. Likewise, others are easily over-aroused by proprioceptive stimuli (e.g., being accidentally pushed in line by a peer, being held back from running into traffic by a parent), setting off a chain of events leading to an emotional meltdown.

Other children with ND-PAE seek out vigorous physical stimulation from falling, jumping, jarring, and colliding with their environment and people in it. For them, early participation in football, gymnastics, wrestling, or vigorous dance can help develop a non-academic skill set, build self-esteem, provide social and recreational involvement, and expend some of their potential energy in a constructive way. For children calmed by rhythmic movement (e.g., movement on swings or roller coasters), they may enjoy participating in various forms of dance such as ballet.

A variety of issues result from the effects of alcohol on the nerves that lead from the brain and spinal cord to voluntary muscles, causing gross and/or fine motor skills deficits. These impairments result in poor handwriting, difficulty coloring or drawing, cutting with scissors and other skills required in elementary school. For some affected individuals, central nervous system damage can lead to coordination problems, spasticity, hyper- or hypo-flexibility, gait disturbances, muscle rigidity, poor tone or hyper tonicity, and other issues that are in the diagnostic realm of "cerebral palsy."

Recent landmark research has shown sensory sensitization to alcohol intoxication. At the time of this writing, researchers have begun understanding that PAE causes children to be overly obsessive and interested in those who are intoxicated if they have used alcohol in the past. A study presented by Dr. Steven L. Youngentob at the 7th National Biennial Conference on Adolescents and Adults with Fetal Alcohol Spectrum Disorder (FASD), "Research on Adolescents and Adults: If Not Now, When?" compared the social behaviors of a mouse with PAE interacting with other mice. If the PAE mouse ever had any alcohol but was not "drunk" currently, they pursued the intoxicated mouse relentlessly with severely maladaptive prosocial behaviors. If they had never had any alcohol, they hardly paid attention to the drunk mouse, similar to a mouse without PAE who was also in the cage. Research shows differences in the chemosensory receptors in the nose and tongue, pointing to alcohol changing taste preferences, smell sensation, and other important areas of sensory functioning.

Figure B-4: Environmental "Cocooning:" Necessary Environmental Considerations for Individuals with ND-PAE

A 4-DIMENSIONAL PERSPECTIVE

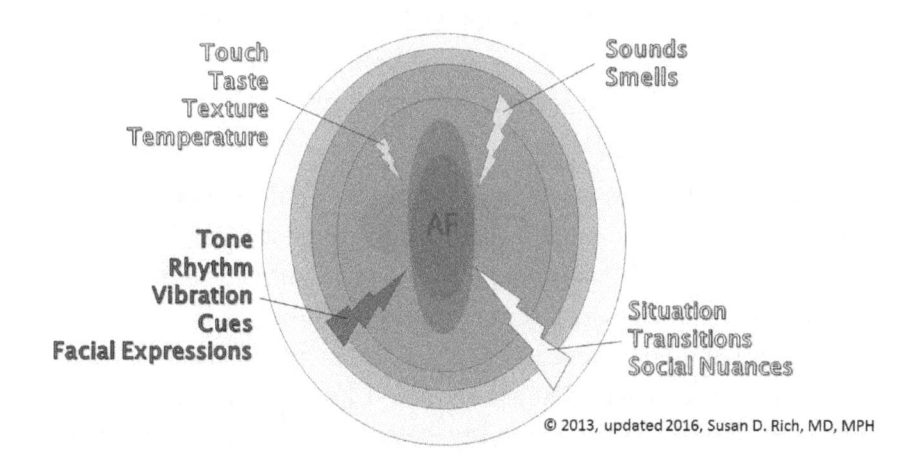

Touch
Taste
Texture
Temperature

Sounds
Smells

Tone
Rhythm
Vibration
Cues
Facial Expressions

Situation
Transitions
Social Nuances

© 2013, updated 2016, Susan D. Rich, MD, MPH

A 4-dimensional model of the experience of a child with sensory hypersensitivities is depicted in Figure B-4. For these children, their stress response becomes "activated" with a number of seemingly innocuous stimuli from the environment – visual, auditory, gustatory, tactile, olfactory, and temperatures. Even perceptions and misperceptions of a person's attitudes or behaviors, nonverbal cues, and voice tone may be misinterpreted as demeaning, judging, off-putting, offensive, or blaming. In that heightened state of arousal, or "fight or flight," the individual with ND-PAE may be more likely to or misinterpret a harmless gesture or comment as a threat.

In addition to becoming agitated when provoked by relatively innocuous stimuli, children with ND-PAE are often more hyper-aroused on a daily basis than typically developing children. Programs like "How your Engine Runs"[172] and other occupational therapy approaches can desensitize the nervous system to environmental stimuli, improve emotional awareness, and self-regulation. Methods of meditation, deep breathing exercises, and simplifying the environment for these children will help improve their functioning by minimizing the insults to their nervous system. The school, parents, coaches, and after school care givers can assist the child or adolescent with emotional regulation by providing appropriate modeling of behaviors, self-control, and reduction of environmental triggers. While there is no "cookie cutter" approach to parenting and caregiving, keeping in mind "wiring deficits" and brain damage that contribute to a child's sensory integration problems will help parents provide cocooning to the delicate nervous system.

---

[172] MS Williams & S Shellenberger. How Does Your Engine Run? ® A Leader's Guide to the Alert Program® for Self-Regulation, 1996. http://www.alertprogram.com/

Appendix C

PARENTING A CHILD WITH ND-PAE

Children aren't born with instruction manuals, and parents don't come with them either. If so, the manual would have to be uniquely written for each person and constantly upgraded – a living, fluid document updated each time experiences and neurodevelopmental wiring bring new attributes and obstacles in their path. – SDR.

Parenting any child is not a task to be taken lightly, particularly one with ND-PAE. The secret is that there is not one "right way" to parent children, no matter what their challenges or strengths. There are some "wrong ways," though … This chapter is meant as a paradigm shift not an instruction manual for parents. The disappointments, challenges, and lovely opportunities of parenting a child with special needs, such as ND-PAE, is captured in the beautiful poetic essay, *Welcome to Holland*, by Emily Perl Kingsley:

Welcome To Holland

By Emily Perl Kingsley

©1987 by Emily Perl Kingsley. All rights reserved. Reprinted by permission of the author.

I am often asked to describe the experience of raising a child with a disability - to try to help people who have not shared that unique experience to understand it, to imagine how it would feel. It's like this … …

When you're going to have a baby, it's like planning a fabulous vacation trip - to Italy. You buy a bunch of guide books and make your wonderful plans. The Coliseum. The Michelangelo David. The gondolas in Venice. You may learn some handy phrases in Italian. It's all very exciting.

After months of eager anticipation, the day finally arrives. You pack your bags and off you go. Several hours later, the plane lands. The flight attendant comes in and says, "Welcome to Holland."

"<u>Holland</u>?!?" you say. "What do you mean Holland?? I signed up for Italy! I'm supposed to be in Italy. All my life I've dreamed of going to Italy."

But there's been a change in the flight plan. They've landed in Holland and there you must stay.

The important thing is that they haven't taken you to a horrible, disgusting, filthy place, full of pestilence, famine and disease. It's just a different place.

So you must go out and buy new guide books. And you must learn a whole new language. And you will meet a whole new group of people you would never have met.

It's just a <u>different</u> place. It's slower-paced than Italy, less flashy than Italy. But after you've been there for a while and you catch your breath, you look around. ... and you begin to notice that Holland has windmills ... and Holland has tulips. Holland even has Rembrandts.

But everyone you know is busy coming and going from Italy ... and they're all bragging about what a wonderful time they had there. And for the rest of your life, you will say "Yes, that's where I was supposed to go. That's what I had planned."

And the pain of that will never, ever, ever, ever go away ... because the loss of that dream is a very very significant loss.

> But ... if you spend your life mourning the fact that you didn't get to Italy, you may never be free to enjoy the very special, the very lovely things ... about Holland.[173]

I frequently describe this poignant parable to parents of children in my practice to help them embrace their child's unique strengths and develop coping strategies for handling the disappointments, sadness, struggles, and problems associated with their day-to-day struggles. As a Sesame Street writer and mother of a now adult son with Down's Syndrome, Ms. Kingsley was able to honor her son's unique abilities and embrace him for the differences rather than chiding him or chastising him for what he couldn't do. It is much easier when a child has a genetic disorder and a parent knows from day 1 that their child has "different abilities." Their parenting is from the approach of "my child is going to be the best s/he can be and we are going to overcome their challenges by capitalizing on their strengths. For children with Down's Syndrome, very often their strengths are in their seeing things in a concrete way, understanding rules rigidly (being able to "call one out" when they are doing something that is against the rules), being super sweet, well-mannered, and even tempered. Their parents and caregivers are enamored by their kindness and applaud each milestone with relief and praise since their expectations of success are much less than "perfection."

That said, if you raise a child with Down's the way that many children with ND-PAE may be parented for the first few years, you will get a similarly dysregulated child. Because so many of them have "faulty wiring," they are difficult to attune and attach with, become anxious over minimal triggers, and avoid situations that stimulate a "fight or flight reaction" by refusing to participate or ruining the evening for the family so that no one else can enjoy the evening if they can't. By the time that many children with ND-PAE come into the lives of their adoptive or foster families, they have experienced years of abuse, neglect, or maltreatment (sometimes by parents who have ND-PAE themselves). They have been locked in a dark closet when they have a 2-year-old temper tantrum (due to "fight or flight" during

---

[173] © 1987, by Emily Perl Kingsley. All Rights Reserved. Reprinted with permission of the author.

an anxiety reaction) and this exacerbates their fear response. Others have been dipped into scalding water as a punishment, or because their cognitively impaired parent does not think to test the water first. Many educated and well-intentioned parents scream at their children with ND-PAE for not knowing or being able to memorize their math facts. Still others become frustrated and demean them for not being able to "initiate" their morning routine (getting up on their own, making their bed, getting dressed, and eating breakfast on time), which makes the parents late for work and further stresses the family system.

In Diane Malbin's book, *Trying Differently Rather than Harder*, she provides a similar approach to parenting and caregiving for children with ND-PAE from the perspective of it being a brain-based condition.[174] She says that we must meet them where they are rather than making them adjust to our way of seeing things. Parenting differently rather than harder and educating differently rather than harder are similar perspectives. We can change ourselves and the environment to adapt to their unique needs – accepting them as they are instead of asking them to adapt to our complex lifestyle. Truly, in parenting children with ND-PAE, we are joining them on their life journey rather than forcing them to keep up with ours.

Parents need not be perfect, just "good enough." For couples who have reproductive health problems or advanced age, the choice to adopt a child or children is often so appealing that they may overlook the trees for the forest. Many are so desperate for children of their own that they miss important clues to the child's underlying neurodevelopmental condition. They may Positive parenting approaches, such as "catching them being good and praising them for it" or working toward tangible rewards are reinforcing of positive behavior in most children, especially those with ND-PAE. Punitive strategies, negative consequences, or withholding pleasurable activities (i.e., recreation, family outings, hobbies, or other extracurricular activities) can be detrimental rather than helpful. Such activities help boost the endorphins – naturally occurring chemicals that help relieve stress, improve mood and concentration, balance sleep-wake

---

[174] Malbin D. *Trying Differently Rather Than Harder: Fetal Alcohol Spectrum Disorders*. Perfect Paperback; May 1, 2002.

(circadian) cycles, and reduce mood outbursts. For children with ADHD, such activities can help the child self-regulate during the school day, at home, and in the evenings before bedtime if they have daily vigorous exercise. More about this and other ways to improve a child's well-being can be found in a chapter on *Therapeutic and Learning Program*.

Parenting Infants with ND-PAE:

A baby who is cared for and nurtured by a consistent, loving caregiver, foster parent, relative, or adoptive parent has a healthier outcome and better prognosis than those left floundering in a biological home with a parent who abuses substances, or in foster care or orphanages. Most parents who adopt children are doing so at some point after birth, so they may not have the advantage of the first few days, weeks, months, or sometimes years of life to see developmental milestones that are missed. It is much easier to recognize and attune to a newborn's rhythms; begin to notice the meanings of their coos and cries for hunger, sleepiness, and other non-verbal cues; and discover the magical ways of communicating and bonding.

It is important for parents and caregivers to accurately perceive the source of the child's distress and to alleviate the upset rather than inadvertently contributing to it by raising the voice or using a demeaning or sarcastic tone. Use the perspective of the parent of a 3 or 4-year-old child for your 10-15-year-old with ND-PAE, who may be unable to accurately perceive his or her own body cues to understand and alleviate the stressor, then gently interpret the feeling for the child – Is she tired? *"It seems like we're tired. Why not take a break for now?"* Is he hungry? *"It's hard to wait for dinner, but we'll have it soon. Would you like some fruit or veggies to snack on?"* Is he scared? *"When the wind blows hard outside, it seems like a storm is coming. Why don't we turn on the weather to see?"* Is she worried? *"Dad hasn't come home yet and it's getting dark. Why don't we call him to see where he is?"*

Parenting children with ND-PAE can seem a bit like it takes twice as long to overcome certain fears, stressors, or other limiting situations. It can seem that your child is "stuck" at a certain developmental age much longer than their peers, who seem to outpace them academically, socially, emotionally, and/or physically. In some ways, it seems as though their body has developed beyond puberty but their mind is stuck in childhood. Often, I tell parents a child may spend 10 years at the emotional developmental age of a 5-year-old and 5 additional years at the social developmental age of an 11-year-old. For this reason, a cocooning environment that incorporates structure and support with nurturing, calming, and protective strategies will enable the metamorphosis of the slowly developing caterpillar (childhood) through the painful process of the pupal stage (puberty) to become the newly emerged, fragile, and delicate-winged butterfly (transitional years). With the right environment, aware and attuned parents, supportive services, and eliminating alcohol and other drugs that will further impair brain development, the young adult can blossom into somewhat of an independent, self-advocating role in the community.

Parents with Alcohol/Other Drug Abuse Problems:

While the ideal situation for most infants is to be cared for by physically and emotionally healthy biological parents, many biological parents are not able to care for their own children safely without significant oversight and supports. The Child Welfare Information Gateway highlights substance abuse problems that interfere with adequate parenting in some of the following ways may need more scaffolding and supports than currently available. Family life for children with one or both parents that abuse drugs or alcohol often can be chaotic and unpredictable. A parent's substance use disorder may affect his or her ability to function effectively in a parental role. Ineffective or inconsistent parenting can be due to the following:

- Physical or mental impairments caused by alcohol or other drugs
- Reduced capacity to respond to a child's cues and needs
- Difficulties regulating emotions and controlling anger and impulsivity
- Disruptions in healthy parent-child attachment
- Spending limited funds on alcohol and drugs rather than food or other household needs

- Spending time seeking out, manufacturing, or using alcohol or other drugs
- Incarceration, which can result in inadequate or inappropriate supervision for children
- Estrangement from family and other social supports.

Children's basic needs—including nutrition, supervision, and nurturing—may go unmet, which can result in neglect. These families often experience a number of other problems—such as mental illness, domestic violence, unemployment, and housing instability—that also affect parenting and contribute to high levels of stress (National Abandoned Infants Assistance Resource Center [AIA], 2012). A parent with a substance abuse disorder may be unable to regulate stress and other emotions, which can lead to impulsive and reactive behavior that may escalate to physical abuse.[175]

During elementary and middle school, academic stressors lead to frustration, emotional distress, and extreme mood dysregulation for children with ND-PAE. Often, after "holding it together" during the school day, they come home and unravel by "vomiting emotionally" onto their families. They may be fatigued by the stress of academics, confused about an assignment, or frustrated with the rigors of homework they do not understand. For some, the vicious cycle of medication withdrawal (i.e., from stimulants they have taken in the morning) can exacerbate such periods of mood dysregulation. Overwhelmed by the stimulation of a busy household just before the dinner hour, these episodes may escalate into physical violence in which they shout, scream, throw things, stomp their feet, slam doors, or hit parents or siblings. Caregivers are often unprepared to provide a suitably cocooning environment, resulting in over stimulating the child's sensitive nervous system, over-reacting to their mood outbursts, or disciplining with harsh or shaming tactics. When the children become frustrated and escalate, the parents will escalate – undermining relational stability. In turn, the essential,

---

[175] The Child Welfare Information Gateway. *Parental Substance Use and the Child Welfare System*, Bulletin for Professionals; October 2014, p. 3 (https://www.childwelfare.gov/pubs/factsheets/parentalsubabuse.cfm).

evolutionarily protective human capacity for bonding is disrupted (see chapter on *Nature versus Nurture – or Both!*).

Family members may feel victimized, traumatized, bullied, intimidated, or battered by these outbursts. Parents who experience their child's volatility often feel helpless to adequately nurture and care for their children. In my own practice, I have worked with families (i.e., siblings and/or parents) who have post-traumatic stress disorder (PTSD) as a result of these rage episodes. Unfortunately, too often parents are disempowered to seek out mental health professionals or not psychologically minded enough to see the importance of therapy for their children let alone themselves. They may feel victimized by the tantrums and outbursts of their child, embarrassed to reach out for help, or feel that mental health issues within the family would somehow stigmatize them within their community. Unlike juvenile onset diabetes, heart disease, cancer, or communicable illness, ND-PAE and associated mental health conditions may be viewed as shameful, socially damning, and reproachful – to the point that parents would avoid seeking help for themselves or their children. Such supports would include early diagnosis, appropriate interventions, academic accommodations, and neuropsychiatric treatment.

During adolescence, impulsive tendencies and peer pressure lead to acting out behaviors, such as sexual promiscuity, substance abuse, conduct problems, and delinquency. These issues invariably affect school performance – either contributing to or a result of academic challenges. For those that can afford years of financially-draining mental health care and medication management for their children, they often are left with young adults who resent having been forced to attend costly therapeutic schools or residential treatment centers. Despite numerous hospitalizations, therapists, psychiatrists, and allied health providers, many families are emotionally scarred by the experience of raising neurodevelopmentally-challenging children. Parents begin to feel demoralized by and ambivalent about the children they had hoped to save. Even loving, well-intentioned yet ill-equipped parents have been forced to disrupt adoptions and foster care placements and some have returned children to their countries of origin.

An intact, healthy parent who is aware of the challenges and "short-circuits" of the neurodevelopmentally-impaired infant is better suited to help develop healthier outcomes than a detached, overwhelmed, distracted, or emotionally unavailable parent with unrealistic expectations of the child's behaviors. Parents with a calm, relaxed temperament will provide appropriate modeling of self-regulation than a parent who becomes unraveled every time the child throws a temper tantrum. In other words, each adoptive, foster care, or biological parent must learn ways of coping with their own sleep deprivation, frustration, anger, and other neuropsychiatric issues in order to overcome the challenges of raising neurodevelopmentally fragile children.

To that end, here are some helpful parenting strategies to cope with neurodevelopmental challenges:

ABC: Always Breathe Calmly

Children with mood dysregulation and autonomic arousal have difficulty regulating their emotional responses and reactions to environmental triggers. When their emotions begin to escalate, if their parent or caregiver escalates too, the child is unable to learn to regulate. However, if the parent is able to remain calm and collected, the child will have an easier time regulating themselves through modeling. For parents with anxiety, depression, bipolar disorder, or other mood issues, it is important to seek help through counseling, medication management and other supportive systems in order to optimize emotional control when the child is unable to control their escalating mood.

Meditation techniques help during such episodes. The technique I teach children and parents is as follows: Imagine your favorite colored balloon inside your head – as though your brain is resting on the coffee table. There are two openings: one to allow air into the balloon at the nostrils and another for dispelling the air from the balloon at the mouth. The point of the exercise is to focus on the breath, letting go of any thoughts, worries, anger, tension, stress or frustration experienced. Do the first steps of breathing a few times while raising your shoulders up to your ears and letting them drop. Begin by taking a deep breath in through the nose, as though filling the balloon in your head with air and collecting all the thoughts and negative emotions. Then exhale out through the mouth as though expelling the negative feelings and thoughts along with the breath. Then continue breathing quietly without making sound – just focus your attention on the breath. This will lower your heart rate and blood pressure, bring your "emotional furnace" down to room temperature, and help calm the child. It's important to practice the technique throughout the day – when waking in the morning, getting ready, driving to work and at your desk at work or doing household chores. It will be easier to remember to breathe through the stressful situation of a child's "meltdown" if you practice doing the technique at other times.

Structure, Supervise, Schedule, and Simplify:

Neurotypical (typically developing) children as well as those with ND-PAE need a routine, predictable schedule, structure, realistic expectations, and nurturing, dependable parents. Often times, parents with busy schedules have caregivers after school and/or on weekends. Limiting unstructured time to supervised play (e.g., allowing the child to play with friends on the playground while supervised by parents) will prevent mishaps and physical aggression toward peers. Explaining the child's neurodevelopmental issues will help care providers and parents of other children intervene appropriately in situations that may trigger outbursts (i.e., distracting and redirecting behaviors rather than shaming or punishing). Having a simpler, more predictable schedule is a relief to a child who is easily over stimulated and has a high degree of anxiety or sensory processing issues. After a long day at school, most children need "down time" at home to decompress and regroup. Scheduling too many activities can be overwhelming and prevent the fragile nervous system from "recharging the batteries" after a long day.

Because most kids with ND-PAE are 2-3 years less mature (dysmature or immature) than their chronological age mates, allowing them the same freedom and independence can be dangerous. One parent of a child adopted from foster care allowed her 6 year old son to go by himself to the neighborhood park across two streets (a few blocks away from her house). She was dumbfounded when she realized he was going into neighbors' houses and taking food or other items. In a sense, she was allowing him to be feral – running the streets without supervision. Most typically developing 6 year old kids are not safe without adult supervision at a park. Kids with ND-PAE at age 13 may be developmentally 8 years old and therefore more at risk of injury, abduction, getting lost, being victimized, or manipulated into misbehavior.

Floor Time, Child-Directed Play, and Free Play:

The late psychiatrist, Dr. Stanley Greenspan, created a timeless method for working with young children – particularly those with autism and other neurodevelopmental issues. It turns out that his approach – "Floor Time"[176] – is successful even for typically developing children. My mentors at Children's National Medical Center, Dr. Jean Thomas and Dr. Irene Chatoor, who ran the program for children from infancy to preschool age taught this approach. Dr. Thomas would often say that spending just 20-30 minutes on the floor everyday with your child will help them laugh, love, learn and listen better. Often, I hear from parents in dual-career households that their children have oppositional or defiant behaviors, being disrespectful and demeaning to the parents. Many times, my advice (echoing my mentors) is to take a bit of time every day if possible to allow the child to choose an activity to do together.

Floor time is not rough and tumble play, it is child-directed play in which the parents allow the child to choose an activity to be played one on one with one of the parents. Rough housing with dad can be fun, but not best to be done at bedtime when the brain needs to be relaxing not hyper aroused. It does not involve the parents talking to each other or talking over the child. The parent is devoting his time to the child, allowing the child to control and direct the play activity. If they choose to play checkers with their own rules, that's okay. If they want to play with cars or dolls, wait till they invite you into the play. Try not to be intrusive (butting in or hogging the toys). Make observational statements with praise like – "Wow! I can see you're a good mommy. You really know how to take care of the baby." Or, "What a good idea! I see how you put the car ramp on top of the books to make it higher." This will

---

[176] Greenspan SI and Wieder S. *The Child with Special Needs: Encouraging Intellectual and Emotional Growth*, Da Capo Press, Inc.; Jan 06, 1998.

help boost their confidence while acknowledging that you are watching and interested in what they're doing. Floor time and child directed play promotes bonding with the parents and self-esteem. Your time is the most precious since they may have so little of it, so the fact that you choose to spend some of it with them will make them feel really special.

Children learn through play activities and also need free play with other children in order to develop social skills, speech and language skills, and develop their identity and sense of self. Children who participate in free play activities as youngsters do better at figuring out their goals in life when it is time to transition to adulthood. Often, children with ND-PAE come from backgrounds or family systems where free play is either not encouraged or not allowed. Many have speech/language problems, anxiety, or other issues that interfere with their ability to interact with typically developing peers without upset or problems. As a parent, helping the child learn how to interact with others through play may be as important as teaching them to brush their teeth or tie their shoes. Some ideas for activities for the following age ranges include:

➢ Infants and toddlers: Sit on a blanket or rug with them on the floor with a few toys, ball, doll in a basket. Allow the youngster to choose something and make comments as above. When they hold out a toy for you to play with, use it in an illustrative way to demonstrate the play to them. There's no need to correct the way they are playing with the doll or toy.
➢ Preschool children: Play dress up, puppet theater, or any type of game they choose. Rough and tumble play may be their preference, but it also may be a distraction from playing in a different way either because they don't have the skill set for imaginative play, have never learned to play that way, or they may be seeking physical contact and affection. If they have been adopted from a neglectful home environment, they may not have ever learned how to play in a creative and inventive way.
➢ School age children: Age appropriate board games, drawing and coloring together, playing with cars together are all fun activities that children in elementary school enjoy doing with a

parent or caregiver. Allowing the child to read with and/or to the parent can build reading skills, boost self-esteem, and improve bonding with the parent.

Chores and Responsibilities:

Beginning at a young age, teaching the child household chores (i.e., making their bed each morning, taking out the trash, emptying the dishwasher, making their breakfast, helping sort laundry, taking their clothes hamper to and from the laundry room) will further strengthen their independence and self-esteem while providing them important life skills. For activities of daily living, many need ongoing supervision to complete most household tasks requiring multiple steps, assistance to organize and clean their room, and help selecting appropriate clothing for the weather. Practicing these areas of "adaptive functioning" will improve their ability to live independently as adults one day, understanding how to care for an apartment or household outside of the confines of the parents' home. In this way, an activity like washing the car together can be done in a child-directed format. Having the child help think of and gather the items needed to wash the car can build their self-esteem, helping them feel good about what they have accomplished when they see the end result. Allowing them to cook with a parent one night a week and join in setting the table or cleaning up afterwards can also provide confidence in being responsible and helping care for themselves.

Another important area of functioning is grocery shopping. Keeping a running list of household and food items needed next to the refrigerator is a good way to help the child learn to efficiently keep track of and take part in grocery shopping. Adding to the list items for meals, ingredients needed for recipes, and other essentials for running the house will help teach the child in a natural way what is necessary to run a household. For parents to take the time to discuss what foods are needed from the grocery store will similarly help build the child's fund of knowledge while involving them in meal planning for the week. At the grocery store, using the list to guide what items go into the grocery cart will assist the child with developing the skill of looking for the items and price comparison. Providing one-on-one guidance in a patient way gives the child time with the parent separate from other siblings as well as minimizing distractions so that they can focus on the task.

Preview, Review & Remind:

Children with ND-PAE are less tolerant of change and transitions in the schedule than their neurotypical counterparts. Many children with ND-PAE lag behind their peers in language development and working memory; therefore, it may be difficult for them to integrate changes in the schedule when told quickly as they are getting out of the car to go to school. If there is a change in the daily routine (e.g., a doctor's appointment) requiring the child to go home by car rather than by bus or the usual route, explaining that to the child the day before a few times and having the child repeat the changes; writing the change down on a paper and having the child read it back or recopy it; then handing the child the note the morning of the appointment after discussing it before school will help the child integrate it into their schedule.

For children who have accompanying mood dysregulation and autonomic arousal along with language and/or executive functioning issues, they may be more likely to have anxiety or an emotional outburst when the schedule changes. In order to prepare the child for the change in the schedule in addition to the "preview and review" strategy above, the parent should provide the school with the recommendation to give the child a verbal reminder and show them the note about the schedule change 30 minutes and 15 minutes before the transition. Have the person (teacher or aide) take the child aside in a quiet place and share the note with them.

Role Play, Replay, and Reframe:

Social situations with typically developing peers may be difficult for the child with ND-PAE to manage, particularly with perceptual deficits, speech/language problems, articulation issues, or intellectual disabilities. As an example, they may be overly sensitive to negativity from peers or take things out of context because they aren't listening carefully to what is said. Although some have difficulty generalizing from one situation to the next, role playing can provide a "script" for working through difficult problems or knowing appropriate ways to handle certain situations. After school each day, replay with them some of the difficult things that happened and try to redirect their perspective in a positive way by reframing what a peer or teacher might have meant that seemed offensive or off-putting at the time. By discussing alternative ways of looking at situations, the child will be able to understand that his way of looking at the situation is not the only way. At times, negativity surrounding a child's experiences of the world reinforces the idea that the child is "bad" rather than her behavior being unwanted. It can be helpful to "replay" situations that ended badly or led to hurt feelings. Saying things like, "Let's start over," or "Can we try this again?" may be difficult for a child with ND-PAE to process. However, "reframing" the scenario by discussing a slightly different outcome will show them that they are capable and able to be successful in similar situations.

Positive Parenting Strategies:

The following positive parenting tips are modified from the CDC's Children's Mental Health Report, listing the following Positive Parenting Tips that you, as a parent, can do to help your child to prepare them for adolescence:

- *Spend time with your child.*
    - *Talk with her about her friends.* Since it is more likely for your child to misinterpret social cues and misunderstand language, reframing social situations in a positive way and providing suggestions for re-interpreting the negative perceptions should be an ongoing part of your daily evening routine.
    - Reflect on the positives in his day, including accomplishments, talents, strengths, and successes. Praise him for the things he is good at and the things you or the teacher noticed that he did well. It could involve being nice to someone, remembering to turn in homework, or bringing home an assignment.

> Discuss what difficulties happened that day as well as challenges she may face. As an example, provide insight into things that happened with a friend – about how what was said may have been misunderstood. Review difficult social interactions with the guidance counselor or teacher to insure your child is not being bullied, misunderstood, or targeted by peers.

- *Be involved with your child's school.* This may include showing up to Parent Teacher Association Meetings, being involved in the Special Needs Parent group, or simply volunteering occasionally in the classroom, cafeteria or at recess. The more involved you are, the more likely you are to notice things that are working well and not working for your child. Be proactive – if you notice things that aren't going well, you'll be able to intervene earlier in their educational process than waiting till they are marginalized in a classroom of peers with poor behaviors distracting them from learning.
  > *Go to school events; meet your child's teachers.* Share information about your child that you notice – whether your child is struggling with math concepts, reading, writing, organization, remembering to complete assignments, forgetting things at school. If the teacher mentions behavior problems, think about ways in which the behavior may be a result of frustration or anxiety about the work being too difficult or not understanding concepts.

- *Encourage your child to join school and community groups, such as a sports team, or to be a volunteer for a charity.* Exercise and good mental health go hand in hand; in fact, daily vigorous exercise improves attention/concentration and mood, while reducing hyperactivity/impulsivity. Often, children sleep better and are able to calm themselves easier when they are active physically. However, coaches can be harsh and players may not be patient with children who are impulsive, hyperactive, or inattentive. The joy of recreational activities are more meaningful for the child if the coaches and players understand the issues your child faces. Getting to know parents of the players and having up front discussions with coaches allows your child to be in an environment where they are included for their abilities rather than excluded for their disabilities. Many of my patients' parents have selected a teenage neighbor to shadow them at sports or scouting activities or even taking them to events or classes at the local recreational center. This can be a great way to have some time to yourself while your child is also enjoying an activity in a safe, structured environment. Often, children with ND-PAE enjoy being around animals, such as dogs and horses. Therapeutic horseback riding, grooming the horses and cleaning the stalls, and volunteering at the local animal shelter or on a farm are all great ways to have interaction with animals in a structured environment. For kids needing one-on-one guidance, the barn staff or humane society can provide guidance and mentoring if they are aware of the special needs of your child.

- *Help your child develop his own sense of right and wrong.* Consequential thinking is not innate for children, adolescents, and adults with ND-PAE. Many children with ND-PAE lack a moral compass – an instinctive conscious knowing of right from wrong. For some, the combination of this deficit, impulsivity, and alexithymia (trouble appreciating feelings) may cause the individual to seem amoral or without conscience. I often see the best outcomes in children whose parents have been able to foster a close tie to religious values at an early age – whether Jewish, Catholic, Protestant, Quaker, Buddhist, Hindu, Muslim, or Unitarian. While such rites of passage as the Bar or Bat Mitzvah, Catholic First Communion and Confirmation, or other religious "coming of age" ceremonies may not be attainable in the strictest sense, a clergy member can guide the child and family through an adapted version of the process. The entire family's involvement in a religious community, the child in Sunday School, and the adolescent in a youth group will provide socialization with families who may be more tolerant of special

needs issues and provide respite for the family in times of need. In that way, religious communities can be a source of strength, support, compassion, and guidance for parents and caregivers.
- *Talk with him about risky things friends might pressure him to do, like smoking or dangerous physical dares.* It is particularly important to begin this conversation at a very early age – stopping to point out people smoking, drinking alcohol, and participating in poor choices. Discussing family members with severe health problems (i.e., dying of emphysema or heart disease) or difficulty in their lives (i.e., losing their license, going to jail, or having a child with ND-PAE) because of their lifestyle choices provides a concrete example that reinforces the life lesson. Explaining that they were exposed before they were born to enough alcohol (and possibly other substances) for 10 people in 10 lifetimes. I often suggest to my patients when they are offered alcohol, cigarettes, or other drugs to say, "No thanks; I'm good. I already had enough."
- *Help your child develop a sense of responsibility—involve your child in household tasks like cleaning and cooking.* As mentioned in the section above on *Chores and Responsibilities*, children who have structured household duties (modified based on the child's functional level) will provide life skills, increase adaptive functioning, and improve self-esteem. Such daily responsibilities teaches important lessons about how to clean the house, fill and empty the dishwasher, prepare healthy meals safely, and other duties that will benefit them as they transition to more independence during young adulthood.
- *Talk with your child about saving and spending money wisely.* Money management is a difficult task for individuals with ND-PAE who may have limited basic math skills, not understand the concept of savings accounts, or appreciate the value of items of food, clothing, or other necessities. Many may require a parent, caregiver, or sibling to assist with their money management during adulthood. If so, consider petitioning for guardianship at some point in their mid-to-late teens.

- *Meet the families of your child's friends.* When children with ND-PAE are placed in special education classes, are hospitalized on a psychiatric unit, stay in a residential treatment center, or participate in another supportive setting, they may be socializing with children who have difficult home environments or challenging life experiences. Because the child/teen with ND-PAE may be gullible, impulsive, and easily duped, they will be more likely to be led and influenced by peer pressure. Often, they are the one "left holding the bag" at a robbery or purse snatching. By meeting your child's friends and their parents, you are at least aware of the background the friend comes from and may be more likely to call their parents if you notice something suspicious happening. The more time you allow the friends to be at your house with your child where you can monitor their interactions and behavior, the better able you are to supervise, redirect, and intervene when appropriate.
- *Talk with your child about respecting others. Encourage her to help people in need. Talk with her about what to do when others are not kind or are disrespectful.* Bad language and behaviors start at home and/or school and should be prevented at an early age by taking the time to speak with your child about what is appropriate and what is not appropriate. It is better to have a well-mannered, well behaved child with special needs than a child who is impolite, unappreciative, disrespectful, and foul-mouthed. Children with milder ND-PAE or more emotional volatility seem to have a more difficult time than the more severely affected child who may be more globally affected in their degree of impairment.
- *Help your child set his own [realistic] goals.* Parents and therapists can provide a concrete list of goals for each day of the week or help the child think about their goal each morning before they go to school. Having a weekly goal written on the inside of the notebook, kept hanging over the bathroom sink, on the refrigerator, and on their bedroom dresser will help reinforce and remind the child about the goal. Personal goal setting should be included in the child's treatment plan. For example, *"I will try*

to raise my hand before calling out in class." "I will breathe in and out slowly when I feel angry." "I will ask for help if I don't know how to do my homework." Reframing future-oriented goals from the context of their strengths and challenges can help them to begin shaping their future vocation or occupation. As an example, noticing the things they are good at or enjoy and pointing out the types of professions that employ someone to carry out those activities can help the child begin to shape a job goal that is realistic and feasible to attain.

- *Encourage him to think about skills and abilities he would like to have and about how to develop them.* This is critical for the child/teen's development of insight; however, often times, a "false bravado" prevents the person with ND-PAE from admitting he lacks the ability to do certain things. They guard their already bruised ego by minimizing their weaknesses and inflating their strengths. Child psychiatrists help parents see the potential in a tween – identifying their strengths and interests, helping the child develop a fund of knowledge in that area, and finding opportunities either in school or by participating in extracurricular activities to enhance their skills.
- *Make clear rules and stick to them.* Consistency in parenting is one of the best ways for a child to learn and remember what is expected of them. A lack of follow through on consequences and inconsistent household rules provide intermittent reinforcement of behaviors – "Pavlovian conditioning," the strongest way to reinforce unwanted behaviors.
  - ➢ *Talk with your child about what you expect from her (behavior) when no adults are present.* Children with ND-PAE can be overwhelmed in a crowded setting or around a number of people. It is easy for them to be distracted by other children or adults and not hear what you are saying. Often, they are embarrassed or easily shamed by displeasing their parents or other adults and can unravel – appearing angry or hostile. It is best to observe and learn from the misbehavior and put systems in place to avoid it the next time by anticipating the triggers for the behavior and modifying

the environment or situation. Preparing your child with ND-PAE ahead of time to prevent specific behaviors may be difficult with just talking about it. Sometimes their working memory may be deficient, they may have trouble generalizing from one situation to the next, or they may act impulsively without thinking about the consequences. Depending on the developmental maturity of the child, role playing and replaying when the situation has passed can assist with reinforcing the behavior that you expect rather than talking through what happened in the immediate aftermath.

> *If you provide reasons for rules, it will help her to know what to do in most situations.* Although it may be difficult for a child with ND-PAE to generalize from one situation to the next, understanding the reasoning and emotional context behind the rule can shed insight to help reinforce the memory. An example is, "I get scared when you run out into the street because I am afraid a car will hit you. That is why we hold hands, stop, and look both ways before we step off the curb."

- *Use discipline to guide and protect your child, instead of punishment to make him feel badly about himself.* Shame is an inevitable consequence of punishment for negative behavior. For children with ND-PAE, redirecting unwanted behaviors and rewarding positive behaviors with praise and attention help reinforce without demeaning or demoralizing the child.
- *When using praise, help your child think about her own accomplishments. Saying "you must be proud of yourself" rather than simply "I'm proud of you" can encourage your child to make good choices when nobody is around to praise her.* The parent or caregiver's voice becomes the "superego" – judge, jury, and prosecutor of the ego. Hearing praise from a parent, including a reference to self-praise, helps the child internalize a positive sense of self rather than a negative attitude about herself.

- *Encourage your child to read every day.* Since reading can be difficult for many children with ND-PAE, request books on tape or E-books from the school or library. Have the child follow along in the hard copy while the book is read to them. Teach your child that knowledge is power and will help them overcome many obstacles. Reading can also offer an escape from stressors and help create new ways of looking at the world around them.
- *Talk with him about his homework.* Homework for children with learning issues, attention problems, or other aspects of ND-PAE can be difficult. Often, the disorganized or inattentive child may not remember the assignment or may not bring home their work. Sitting still long enough to complete homework after a long day at school can also be challenging. Sometimes parents need to suggest that the child's Individualized Educational Plan include time during the school day to complete the work with an aide or that they stay afterschool for assistance. It is more important that a child with ND-PAE have time to decompress from a tough day at school than spend hours sitting at the dining room table practicing math problems or doing busy work they could have done during the day. Spending that time engaging with the child in a child-centered fun activity, such as a board game or visiting the park is healthier and more meaningful than fighting over homework completion.
- *Be affectionate and honest with your child, and do things together as a family.* Affection doesn't have to be physical hugging or holding, although even kids with sensory issues need to feel loved, nurtured, and cared for. Spend quality time with your child, expressing how your day was and asking them about their day. When they ask if you're okay if you're upset, let them know that you are sad or disappointed about something that happened instead of brushing off their concern by saying, "Honey, Mommy is fine." Acknowledge their being able to notice your feelings. Honesty about your own feelings can help your child with ND-PAE develop a better understanding of their own emotions as well as accurately perceiving emotions of others.

Babies are hard-wired to perceive facial expressions, tone of voice, and other non-verbal cues from an early age. In a neurodevelopmentally impaired child, these innate abilities can be altered in such a way that they misperceive, over-react to, or become confused by these subtle hints about meaning of spoken language. Infants who are institutionalized from an early age may have additional difficulty accurately perceiving social cues or non-verbal cues because they were unable to attune to a caregiver long enough to appreciate the unspoken meaning implied in gestures and facial expressions. Children who have multiple care providers as infants may also have difficulty accurately reading such social cues, especially if their primary caregiver is giving them mixed messages.

- *Talk with your child about the normal physical and emotional changes of puberty.* Several studies show that puberty is occurring on average two years earlier in American boys and one year earlier in American girls compared with decades ago. While increased sedentary lifestyles, "oversized" meals, junk food, growth hormone and environmental estrogens in the food supply have all been blamed on the obesity epidemic and precocious puberty, brain changes associated with prenatal alcohol exposure is also a culprit. Although children with ND-PAE can be emotionally and socially dysmature or immature, their development can be affected by hormonal differences (i.e., neuroendocrine dysfunction), though not necessarily causing precocious puberty.[177] They may have weight gain on medications to assist with mood control (i.e., antidepressants or antipsychotics) that lead to early breast development or gynecomastia (male breast development). Being able to discuss these changes with your child with the assistance of a medical professional, as well as the normal developmental changes of menarche in girls and puberty in boys will help the child through these changes easier.[178]

Highlighted Case

---

[177] Carter RC, Jacobson JL, Dodge NC, Granger DA, and Jacobson SJ. Effects of Prenatal Alcohol Exposure on Testosterone and Pubertal Development. *Alcoholism: Clinical and Experimental Research;* Volume 38, Issue 6, pages 1671–1679, June 2014.

[178] See Minnesota Organization on Fetal Alcohol Syndrome website: http://www.mofas.org/resource/precocious-puberty/

One young man with full FAS, "Thomas," referred by a diagnostic clinic in 6th grade and in my practice for several years (now in his early 20s), worked as a bagger in a grocery store from age 15 through completion of his modified diploma. While he lacked social skills, his mother talked to him about his interactions with the cashier he was frequently paired with, reframing and making suggestions about negative ways in which they each "pushed each other's buttons." Thomas was often disgusted by the way the cashier spoke to him or to customers in a haughty or disrespectful tone. His mother redirected his understanding by helping him see that she may be having a bad day, experiencing stress at home, or may not enjoy her low wage job. His "black and white" thinking allowed us to capitalize on issues of right and wrong by identifying words that people may misconstrue and judge him for.

Several years ago, Thomas and I developed a list of "red words" that he should stop using – foul language that he heard at school or from people in the community. Such words had led to misunderstandings between him and classmates or teachers, thinking he was being disrespectful or "trash talking." We discussed ways of catching himself in the act of using the word and thinking of an alternative word to use instead. Over time, by role playing and replaying the scenarios and having his parents reinforce his not using the words, Thomas has been able to inhibit his use of such language. Like Thomas, children hear words that are inappropriate at home or school and become tolerant to them in such a way to use them all the time. This creates an aura of "gutter language" for a child that may already struggle with speech and language issues.

With other patients and my own children, identifying these "red words" has helped eliminate inappropriate language. These are very strong words that make people believe you're not as smart as you are if you say them, or words that start wars. I use the example of saying "Bomb" in the airport – a word that leads to the person being escorted to a locked room and stripped searched along with anyone traveling with them. Anything that sounds like "Oh, my g--!!" or "S--- Up!" or "I H--- You!" (the "H" word) or "I'll K--- You!" (the "K" word) are "red words" that are not to be used starting as soon as the child can talk. I let patients know that certain phrases will guarantee a ticket straight to the psychiatric ward ("I want to k--- myself!" or "I want to k--- you!"). Parents should also not use those words (or words like "freaking," "stupid," "dumb"), even in conversations with other adults. Teaching these words and phrases from a very young age to children when they first encounter the word (or use it) enables the parent to prevent further development of worse vocabulary. The idea that words have power and meaning can develop the child's understanding of their influence over others with language.

Thomas always enjoyed taking things apart as a youngster to see how they worked. He and his father, who is an engineer, would go to yard sales and swap-meets to get broken TV's, electronics, radios, pumps, and other machines to take apart and put back together. Although he did not have a vocational program at his school, his father increased his fund of knowledge and ability to "fix things" by working together on those projects. Repairing lamps, clocks, toaster ovens, and transistor radios gave the patient a sense of accomplishment, job training skills, and invaluable time with his father. He is now attending community college classes, having continued in the high school transition program until he was 21 years old. He drives himself to and from school, has never been involved in smoking cigarettes or using alcohol/other drugs. Despite significant learning and attention problems as well as executive functioning issues, he has good self-esteem, sees himself as more competent than most typically developing kids his age (who don't know how to fix things), and maintains a great relationship with his parents. When he was in for his last appointment, we discussed the three online community college classes he was taking in psychology and social work. As of the mid-term grades, he had A's in all three classes.

Transitional Age Youth:

It is not simply rearing in a safe, nurturing home environment that will enable adolescents with this condition to successfully transition into adulthood. Developmentally, they are much younger than their chronological ages, so insisting that they become independent, functional, self-sufficient members of society at the end of adolescence is not always realistic. They need ongoing support to find vocational opportunities, to attend college or technical/trade school, and to lead meaningful, productive lives. Using their strengths and developing skills in an area of interest such as building/construction, landscaping, nature, forestry, animal caretaking, or clerical work when they are in middle school and high school will help provide after school enrichment in addition to vocational programming. By giving them apprenticeship programs and volunteer experiences at a relatively early age, they will begin to develop self-confidence and experience, job skills, vocational etiquette (e.g., use of proper dress, language, social interactions), and time management. Because all of these life skills will take longer for them to develop than typically developing kids, it is important to begin early so that they will have the skills they need by the time they graduate high school. Even if they go on to college or technical school, these adolescent experiences will provide the beginnings of a work ethic, responsibility, and involvement in an after school activity to hopefully keep them occupied and out of trouble.

The transitional years are perhaps the most difficult and the most critical in the lives of individuals with ND-PAE. This is the time when their peers are beginning to leave home to find their way in the world, launching off to college, and/or working to support themselves. Our society now recognizes the increased window of adolescence well into the mid to late 20s, with insurance plans allowing parents to keep young adults on their plans until age 26 and the scientific community acknowledging that typical brain development continues beyond age 25 or 26. Given that the brain wiring deficits of ND-PAE lead to an extended period of brain maturation, re-architecture and growth, transitioning into adulthood may extend into the early to mid-30s.

Pre-DSM-5 psychiatrists had managed patients with prenatal-induced encephalopathy based on symptomatic behaviors, leading to a plethora of diagnoses and medication "cocktails" by adolescence and adulthood. Undiagnosed or misdiagnosed adolescent and young adult patients become demoralized, socially disenfranchised, despondent and demeaned by a system of care that misunderstands their behaviors as willful rather than maladaptive, misinterprets their moods as "bipolar" rather than mis-wiring of their "fight or flight" reaction, and overmedicates rather than changing the environment to accommodate their encephalopathy. In turn, failure to adequately diagnose and treat the underlying neurodevelopmental problems coupled with a lack of appropriate parent guidance, academic supports, social skills development, and medication frequently lead to negative outcomes such as school failure, violent and aggressive behavior, and problems transitioning to a meaningful young adulthood. Impairment in neurocognitive abilities, emotional regulation, sensory/motor skills, and social communication often lead to chronic disruptions in developmental milestones, family life, school performance, vocational experiences, relationships and other areas of adaptive functioning. Such deficits contribute to worsening of symptoms and poor outcomes (e.g., disruptions in jobs or vocational training, frequent hospitalizations, institutionalization, and/or adjudication). By the time that many individuals have reached adulthood, they have been mismanaged to the point that their maladaptive behaviors have landed them in jail, prison, prostitution rings, homeless, or on death row.

Though the CDC estimates that as many as 1 in 20 (2-5%) school age children have ND-PAE, episodic waves of epidemic use of alcohol during pregnancy has led to spikes in the number of affected individuals. With a 400% relative increase in moderate to heavy maternal alcohol use from 1991 to 1995, these young adults are now entering young adulthood. The hope of a brain-based, neurodevelopmental formulation provided by inclusion of FASD in DSM-5 is that clinicians will recognize the significance of encephalopathy (brain damage) and the importance of a structured, stable, supportive environment to optimize the individual with ND-PAE's adaptive functions. An understanding of the neurodevelopmental nature of a patient's emotional dysregulation, social communication problems, neurocognitive deficits, and sensory/motor issues can help scaffold the environment as the individual passes through puberty and young adulthood.

It is my strong belief that accurate diagnosis and intervention can prevent secondary disabilities (academic failure, criminal behaviors, etc.) and avoid over-medication or inappropriate treatment. Increased understanding of ND-PAE, improved coordination of care, integrating farm animal therapy into clinical practice, and psychopharmacologic treatments hold promise for improved prognosis for affected individuals.

Afterword

Today, Monday, January 25, 2016 is the third day of winter snow storm, Jonas, dubbed "Snowzilla" by those of us living through the blizzard conditions. The storm hit the East Coast with such fervor that the schools were closed on Friday before the first snow flake fell in Greater Washington, DC, an area of the country where a dusting usually leads to "Black Friday" conditions at grocery stores. Little did I know that the storm would bring a wakeup call, shifting my own paradigm about boundaries and personal safety.

After three days ensuring my chickens and therapy goats were warm and well cared for, watching Star Wars and Indiana Jones movies snuggled by the fire with my kids, and finally taking down the Christmas tree, I was ready for a warm shower yesterday afternoon. My daughter came into the bathroom midway through and said, "Mommy, Nathan is here." I called back over the shower, "Nathan, who, Sweetie?" and she answered, "You know, Mommy, your patient." My mind immediately thought that his mother may have dropped off the 20-something young man to help clear the snow from my driveway. She and her partner had adopted several children, including Nathan and his siblings, from foster care and always taught them to take care of other people in times of need. Although our street was still impassable, I believed the benevolent family might have found a way to get here – two towns over, to help me clear my driveway. Nonetheless, I sent her back to ask why he came and how he got here. Just as she left the bathroom, I realized how vulnerable she and my son were with no one to protect them if Nathan came in the same "Hulk-like" rage that landed him in the hospital just before the

Christmas holidays. At that point, my name was on a very short list of people he intended to kill.

Quickly drying off and pulling on some clothes, I made my way into the living room finding Nathan having a friendly chat with my children as though he were a 10 or 12-year-old neighbor coming in out of the cold to get hot chocolate. I greeted him and asked how he got there and what he was up to, given the state of emergency our area was in. He said he and his mom had "gotten into it" and he rode his bike to my house because he needed to get away. He rode his bike on nearly impossible roads – 12.1 miles in the aftermath of a blizzard. What I realized then is how desperate he must have been and how sad that I am the only person he could think of to help in the situation. After piecing what had happened together from a phone conversation with his mother and talking to him while I made dinner, it seemed they had argued over "who ate two pop tarts?" which sent him into an angry rage because they both believed their side of the argument and were not budging. The frustration sent Nathan into a rage when his amygdala was pushed into "fight or flight," and when his other mother suggested that he go outside to cool off, he grabbed a coloring book and a few other items and put them into his back pack and headed out the door. Unbeknownst to them, he carried his mountain bike to the nearest main thoroughfare and made his way to my house.

On one hand, I might be flattered that Nathan thought of me helping him during such a crisis. However, I am a psychiatrist not a martyr, a friend, or Mother Teresa. I have my own children's health and safety to think about and my own life to keep intact so that I can provide for them and care for my other patients. In this situation, I realized that I needed to regroup, to simplify, and to reframe my thinking. Like many others with ND-PAE, Nathan is often incredibly impulsive, lacks judgement, can be violent, misperceives innocuous social slights, and has limitations in consequential thinking. If he would drive through impassible road conditions to get to my house, what might he do in good weather? Recognizing the improbability of him making it back home before dark and of my being unable to get my car out of the garage let alone out into the street, I decided to rethink his mother's suggestion to "send him home the same way he got there." After discussing the limited options with his mother, we decided that the best choice would be to call 911 to have them help me find a way to get him home. Of course, it took nearly two hours for the police to arrive with the road conditions. In the meantime, I enlisted Nathan's help to assist with beginning to clear my driveway then to help me make spaghetti sauce for dinner. He was extremely helpful, giving me tips his Italian mother uses during her weekly sauce-making. By the time the police officers arrived, we were all sitting having dinner in the dining room as though it were as usual an evening as any other.

Explaining to Nathan the risks involved in riding his bike on deserted roads in a state of emergency during the aftermath of the heaviest snowfall in decades was a poor use of the police officers' time. Nonetheless, we agreed to emphasize the risk involved for Nathan in biking 12 miles in the snow, for my children who were unaware that he could have been a threat, and for me in being unable to get him home during a blizzard. We agreed to leave his bike and helmet at my house until the snow cleared and for the police to drive him home for his safety, not because he was in trouble with the law. Although he had unwittingly astonished and frightened me to no end, I was not yet aware of the profound "reality check" that his unexpected visit on that cold, snowy Sunday would have on my perspectives about these kids. I had often heard parents speak of the times when their children or adolescents had taken risks without an awareness of safety issues or the harm that might come to themselves or someone else. I understood later that Nathan innocently thought that I would somehow help his situation by solving the problem with the pop tarts. Since his mom had taken his phone away, he couldn't call or text me. In the rage of "fight or flight," he fled the house without having any idea of where he might go. As he peddled, his mind recognized the road and unconsciously took him to the place that he knew had been helpful in the past – my office, which is also my home. Later, when we discussed his thought processes, he said he thought my road and driveway would be cleared and that I would talk to his parents so they would understand about the pop tarts.

Despite my passion and convictions that these adolescents and adults can be helped, I also realize that I cannot help everyone. More importantly, I am unwilling to put my own children at risk in the event someone like Nathan decides to drop by unexpectedly. Although I am willing to venture through the halls of death row to interview and evaluate individuals with prenatal alcohol exposure, I have to draw the line when it comes to my family. I have made up my mind to focus my efforts on younger children – identifying their needs at an earlier age and helping parents cocoon them with the right supports and nurturing environment as they grow through adolescence into young adulthood. They have potential if given the right field to grow in ... together, they can create a transformed world by showing society what they are capable of in a positive way. But they need support, encouragement, healthy connections, healthy role models, and time to heal their developing nervous system.

While writing this final section of my book, the Centers for Disease Control and Prevention has just come out with a publication warning childbearing age alcohol consumers to use birth control in order to prevent Fetal Alcohol Syndrome. This important paradigm shift is critical to prevent a condition that occurs at a time before most previous public health warnings about alcohol exposure take effect (i.e., prior to pregnancy recognition). At the same time, the most basic of information – "why contracept?" was missing from the lead-in to their advisory – because alcohol can cause problems as early as the first 3 weeks after conception, with Fetal Alcohol Syndrome resulting from as little as 4-5 drinks in one sitting. Therefore, alcohol consumers should use a reliable contraceptive until planning to be pregnant, then avoid alcohol after stopping the contraceptive.

None of us are infallible in our efforts to diagnose, treat, and prevent ND-PAE. It will take each of us together, flapping our proverbial wings in order to stem the tide in this epidemic. The butterfly effect in quantum mechanics holds true for those of us with the courage, passion, conviction, and intellect to raise awareness to spite, despite, and in spite of an industry that counters science with advertising. Not unlike the tobacco industry before widespread public recognition and unrest about the long term health risk of second hand smoke, the alcohol industry has duped our inebriated culture through overt advertising and marketing. "Big alcohol" spends more on sexy commercials to entice childbearing age alcohol consumers to drink, party, play, and "date" while under the influence – that even my daughter at age 10 could understand the unethical principle – marketing to people who would have sex under the influence and make more brain damaged babies.

For those who believe that terms such as zealot, maverick, provocative, audacious, impractical, and unrealistic would offend me, I welcome the opportunity to explain my position. I do not apologize for hope, optimism, and change. I was trained as a change agent in public health and believe as a physician that we can and should help those who may be less fortunate or ill informed. Let's put an end to brain damage caused by an industry that produces a product with very little value other than as a preservative or solvent. Since alcohol influences libido while at the same time causing brain damage in the babies that would be produced by the consequences of unrestrained libido, the solution is simple – require a condom to be affixed to every bottle of beer, wine, or liquor sold. Perhaps then, the message will be clearer to the public than a tiny, nearly illegible, inconspicuous label or mixed messages from the media.

The children are our future and we must protect them (and their brains) at all cost.

Appendix D

These are only a sampling of resources for professionals, families and people living with ND-PAE.

PRINTABLE RESOURCES:

FASD Prevention – From NOFAS:

http://www.nofas.org/wp-content/uploads/2014/05/Facts-prevention.pdf

What young people should know – From NOFAS:

http://www.nofas.org/wp-content/uploads/2014/05/Fact-sheet-young-people.pdf

What school systems should know – From NOFAS:

http://www.nofas.org/wp-content/uploads/2012/05/school-prevention.pdf

A Step by Step Guide for Primary Care Practices – CDC Resource

https://bettersafethansorryproject.com/category/cdc-resources/

OTHER WEBSITES:

http://www.fasworld.com/fasworld-toronto/

http://americanpregnancy.org/pregnancy-complications/fetal-alcohol-syndrome/

http://www.cdc.gov/ncbddd/fasd/

http://www.cdc.gov/preconception/index.html

http://www.cdc.gov/ncbddd/fasd/training.html

http://www.niaaa.nih.gov/research/major-initiatives/fetal-alcohol-spectrum-disorders

http://www.uaa.alaska.edu/arcticfasdrtc/Resources/Professional/alaskaresources.cfm

http://www.healthybrainsforchildren.org/

https://thenationalcampaign.org/

ARTICLES WRITTEN BY DR. RICH AND COLLEAGUES:

Consumer Protection and the Industry's Duty to Warn: http://link.springer.com/chapter/10.1007%2F978-3-319-20866-4_3

ND-PAE:

http://www.casaforchildren.org/site/c.mtJSJ7MPIsE/b.8968413/k.9568/JP3_Rich_Brown.htm

FASD and Autism:

http://www.intechopen.com/books/recent-advances-in-autism-spectrum-disorders-volume-i/clinical-implications-of-a-link-between-fetal-alcohol-spectrum-disorders-fasd-and-autism-or-asperge

NOFAS UK:

http://www.nofas-uk.org/PDF/NOFAS-UK%20Fetal%20Alcohol%20Forum%202010.pdf (pp. 34-37)

Psych News 2013:

http://psychnews.psychiatryonline.org/doi/full/10.1176/appi.pn.2013.11a18

Psych News 2005:

http://psychnews.psychiatryonline.org/doi/full/10.1176/pn.40.9.00400012

Reference Books:

Addressing Fetal Alcohol Spectrum Disorders (FASD), by SAMHSA.

Reviews alcohol screening tools and interventions for use with pregnant women and women of childbearing age to prevent fetal alcohol spectrum disorders (FASD). Also, outlines methods for identifying individuals with FASD and modifying treatment accordingly.

Alcohol and Neurobiology: Brain Development and Hormone Regulation, edited by Ronald R. Watson.

This information will assist the researcher, clinician, and student in comprehending the complex changes caused by direct and indirect effects of single drugs at the cellular level.

Cheers! Here's to the Baby! A Birth Mother's Discovery of Fetal Alcohol Syndrome, by Linda Belle La Fever.

This book reveals the struggles and heartbreaks of a loving single mother who unknowingly made a terrible mistake that forever altered the potential of her youngest son. Danny, now a young adult, was diagnosed when he was five years old.

Damaged Angels, by Bonnie Buxton.

Part heartfelt memoir, part practical guide, Damaged Angels recounts Bonnie Buxton's struggles to raise an adopted daughter whom she didn't realize was afflicted with fetal alcohol disorder. Her book also offers guidance to parents who have children with FASD.

Fit to be Tied: Sterilization and Reproductive Rights in America, 1950-1980, by Rebecca M. Kluchin.

Rebecca Kluchin impressively navigates a critical period in the history of reproductive health in America. The book is very innovative in a subtle and understated way: Kluchin is one of the first historians of gender and medicine to provide a sophisticated framework for mapping the sterilization practices of the pre-World War II period into the post-Roe V. Wade culture.

Forfeiting All Sanity, by Jennifer Poss Taylor.

Jennifer Poss Taylor shares her family's experience with FAS and the perseverance, sense of humor, and love that daily overcome its effects. Taylor's honesty and personal insight will capture readers as she describes the daily challenges of raising a child with special needs. Every parent will be touched by this story.

Ghosts from the Nursery: Tracing the Roots of Violence, by Robin Karr-Morse and Meredith S.Wiley.

As the nation becomes alarmed by reports in the media of the growing wave of violent children, *Ghosts from the Nursery* presents startling new evidence that links aggressive and violent behavior to the effects of abuse and neglect on the infant brain.

Karli and the Star of the Week, developed by NOFAS.

Developed and incorporated into a lesson that encourages students to understand and be accepting of others' abilities and disabilities. The module includes the storybook, lesson plan, and a CD-ROM with teacher background information.

Message in a Bottle: The Making of Fetal Alcohol Syndrome, by Janet Golden.

In *Message in a Bottle*, Janet Golden charts the course of Fetal Alcohol Syndrome through the courts, media, medical establishment, and public imagination.

My Invisible World: Life with my Brother, His Disability, & His Service Dog, by Morasha R. Winokur.

Eleven-year-old Morasha shares her in-depth and personal life story of what it feels like to be the sister of a sibling that struggles with FASDs. *My Invisible World* is an incredible inside account of the daily issues that arise for a child that deals with an invisible brain injury.

Our Fascinating Journey: Keys to Brain Potential Along the Path of Prenatal Brain Injury, by Jodee Kulp.

Our FAScinating Journey will introduce readers to another winding path in working with prenatally exposed children. Jodee illuminates this path with lights that shine the hope of possibilities for these special kids.

Supporting Caregivers of Children with Fetal Alcohol Spectrum Disorders, by Anne Hedelius.

This paper addresses the challenges of creating and sustaining a support network for parents of children with an FASD.

The Broken Cord, by Michael Dorris.

The controversial national bestseller that received unprecedented media attention, sparked the nation's interest in the plight of children with Fetal Alcohol Syndrome, and touched a nerve in all of us. Winner of the 1989 National Book Critics Circle Award.

The Challenge of Fetal Alcohol Syndrome: Overcoming Secondary Disabilities, edited by Ann Streissguth and Jonathan Kanter.

In the first book of its kind, experts describe how to help people with Fetal Alcohol Syndrome. A summary of recent findings and recommendations is presented by the team who conducted the largest study ever done on people of all ages with Fetal Alcohol Syndrome and Fetal Alcohol Effects.

The Fatal Link, by Jody Allen Crowe.

This book reveals the undeniable connection between school shooters and their mother's alcoholic behaviors. The author, Jody Allen Crowe, is uniquely positioned to identify this never before seen profile of school shooters. He is an educator who gained his knowledge working in the epicenters of violence and abnormal behaviors.

The Insanity Offense: How Americas Failure to Treat the Seriously Mentally Ill Endangers its Citizens, by E. Fuller Torrey.

E. Fuller Torrey, the author of the definitive guides to schizophrenia and manic depression, chronicles a disastrous swing in the balance of civil rights that has resulted in numerous violent episodes and left a vulnerable population of mentally ill people homeless and victimized.

The Long Way to Simple, by Stephen Neafcy.

Steve, an adult with FAS, offers a fresh perspective into FASD from the affected individual's experience. This easy read offers insights for parents and professionals and encouragement for persons who suffer brain damage from prenatal exposure to alcohol. Visit Steve's web site FASFlight.com.

Trying Differently Rather Than Harder, Second Ed.; by Diane Malbin.

This book provides a readable, narrative discussion of the neurobehavioral approach for working effectively with children, adolescents and adults with FASD.

Understanding Fetal Alcohol Spectrum Disorder: A Guide to FASD for Parents, Careers, and Professionals, by Maria Catterick and Liam Curran.

This is the essential guide to FASD – the most common non-genetic learning disability, which is caused by alcohol consumption during pregnancy. It explains how FASD affects individuals at different stages of their lives, how you can identify it, and gives advice on how to support children, young people and adults with FASD.

When Rain Hurts: An Adoptive Mother's Journey with Fetal Alcohol Syndrome, by Mary Evelyn Greene.

When Rain Hurts is the story of one mother's quest to find a magical path of healing and forgiveness for her son, a boy damaged by the double whammy of prenatal alcohol abuse and the stark rigors of Russian orphanage life.

About the Author

Susan D. Rich, MD, MPH, DFAPA is a psychiatrist specializing in the diagnosis, treatment, and prevention of neurodevelopmental disorder associated with prenatal alcohol exposure. She has worked since 1993 to raise awareness among medical professionals, policy makers, and community leaders about this prevalent, preventable condition. Her nonprofit, 7th Generation Foundation, Inc., provides support and social opportunities for youth with ND-PAE.